XIII Brazilian Materials Research Society Meeting 2014

IOP Conference Series: Materials Science and Engineering
Volume 97

Joao Pessoa, Brazil
28 September - 2 October 2014

ISBN: 978-1-5108-1634-3
ISSN: 1757-8981

Printed from e-media with permission by:

Curran Associates, Inc.
57 Morehouse Lane
Red Hook, NY 12571

Some format issues inherent in the e-media version may also appear in this print version.

Printed by Curran Associates, Inc. (2016)

For permission requests, please contact the Institute of Physics
at the address below.

Institute of Physics
Dirac House, Temple Back
Bristol BS1 6BE UK

Phone: 44 1 17 929 7481
Fax: 44 1 17 920 0979

techtracking@iop.org

Additional copies of this publication are available from:

Curran Associates, Inc.
57 Morehouse Lane
Red Hook, NY 12571 USA
Phone: 845-758-0400
Fax: 845-758-2634
Email: curran@proceedings.com
Web: www.proceedings.com

Table of contents

Volume 97

Brazil MRS Meeting 2014
28 September to 2 October 2014, João Pessoa, Paraíba, Brazil

Accepted papers received: 9 October 2015
Published online: 11 November 2015

Preface

Papers

IOP Conf. Series: Materials Science and Engineering **97** (2015) 011001 doi:10.1088/1757-899X/97/1/011001

Preface

The annual meetings, organized by the Brazilian materials research society – B-MRS, are amongst the most import discussion forums in the area of materials science and engineering in Brazil, with a growing interest from the national and international scientific society. In the last 4 years, more than 1,500 participants have attended the B-MRS meetings, promoting an auspicious environment for presentation and discussion of scientific and technological works in the materials science area.

The XIII Brazilian Materials Research Society Meeting was held from 28 September to 02 October, 2014, in João Pessoa, PB, Brazil. The Meeting congregated more than 1650 participants from the whole of Brazil and from 28 other countries. More than 2100 abstracts were accepted for presentation, distributed along 19 Symposia following the format used in traditional meetings of Materials Research Societies. These involved topics such as: synthesis of new materials, computer simulations, optical, magnetic and electronic properties, traditional materials as clays and cements, advanced metals, carbon and graphene nanostructures, nanomaterials for nanostructures, energy storage systems, composites, surface engineering and others. A novelty was a symposium dedicated to innovation and technology transfer in materials research. The program also included 7 Plenary Lectures presented by internationally renowned researchers: Alberto Salleo from Stanford University, United States of America; Roberto Dovesi from Universita' degli Studi di Torino, Italy; Luís Antonio F. M. Dias Carlos from Universidade de Aveiro, Portugal; Jean Marie Dubois from Institut Jean-Lamour, France; Sir Colin Humphreys from University of Cambridge, England; Karl Leo from Technische Universität Dresden, Germany; Robert Chang from Northwestern University, Evanston, United States of America.

The numbers of participants in the B-MRS meetings have been growing continuously, and in this meeting we had almost 2200 presentations distributed in plenaries, invited, oral presentations and posters. From these presentations, each symposium chose the best oral and poster presentations presented by students, which were awarded with the "Bernhard Gross Award". Some papers of these awarded works are presented in this proceeding.

Conference photographs

Partial view of the convention center during the meeting

IOP Conf. Series: Materials Science and Engineering **97** (2015) 011001 doi:10.1088/1757-899X/97/1/011001

Opening ceremony of the XIII B-MRS Meeting. From left to right: Jackson Lima, Claudio Furtado, Ieda Garcia, Roberto Faria e José A. Varela

Presentation of the lecture "Memorial Joaquim Costa Ribeiro", by Prof. José Arana Varela during the Opening Cerimony

1st Meeting of the University Chapters

IOP Conf. Series: Materials Science and Engineering **97** (2015) 011001 doi:10.1088/1757-899X/97/1/011001

Discussions during the poster session

One of the simultaneous oral presentations of the 19 symposia in 14 different rooms

Visit to the 30 exhibitors during the Meeting

IOP Conf. Series: Materials Science and Engineering **97** (2015) 011001 doi:10.1088/1757-899X/97/1/011001

Winners of the best oral and poster presentation of each symposium

Symposium A - Poster

Symposium B - Oral

Symposium C - Poster

Symposium D – Oral

Symposium D – Poster

Symposia E/G/P – Oral

Symposium F - Oral

Symposium G - Poster

Symposium H – Oral and Poster

Symposium K – Oral

Symposium K – Poster

Symposium L – Oral

IOP Conf. Series: Materials Science and Engineering **97** (2015) 011001 doi:10.1088/1757-899X/97/1/011001

Symposium L – Poster

Symposium M – Oral

Symposium M – Poster

Symposium N - Poster

Symposium O – Poster

Symposium P – Poster

Symposium S - Poster

Sponsors

Realization

SBPMat Brazil-MRS

Universidade Federal da Paraíba

Support

CNPq CAPES FAPEMIG FAPESP

IOP Conf. Series: Materials Science and Engineering **97** (2015) 011001 doi:10.1088/1757-899X/97/1/011001

Sponsorship

Elizabeth Cimentos

FEI

SHIMADZU Excellence in Science

Exhibitors

Agilent Technologies

altmann

ANACOM CIENTÍFICA

Anton Paar

arotec

US Bio solutions

Dafratec

BRUKER

C K

Leica

dp UNION

EDWARDS

FORTELAB

HORIBA

Instituto Nacional de Engenharia de Superfícies

INSTRON EMIC

Instrutécnica

IOP Publishing

JEOL

LABCONTROL

Metrohm

micromeritics Tecnologia de Particulas

M&M Vácuo

MTS

ZEISS

NETZSCH

OXFORD INSTRUMENTS The Business of Science

QuantumDesign

ReoTerm INSTRUMENTOS CIENTÍFICOS

SHIMADZU Excellence in Science

List of committees

Conference Chairs

Iêda Maria Garcia dos Santos
Universidade Federal da Paraíba, UFPB, João Pessoa, PB – Brazil
Severino Jackson Guedes de Lima
Universidade Federal da Paraíba, UFPB, João Pessoa, PB - Brazil

Program Committee

Márcia Rejane Santos Silva (UFPB)
Dannielly Silva de Melo (UFPB)
Fernando Volpi de Oliveira (UFPB)

Awarded presentations

Simposium A - Functional hybrid interfaces: from characterization to applications
Poster presentation: Oigres Daniel Bernardinelli, Igor Polikarpov, Eduardo Ribeiro de Azevedo; Information on crystallinity index of sugarcane biomass submitted to a chemical and enzymatic treatment via solid-state nuclear magnetic resonance (SSNMR)

Simposium B - Ceramic and metallic materials obtained by wet-chemical methods
Oral presentation: Thiago Marinho Duarte, Luzia M. C. Honório, Juliana K. D. Souza, Arnayra S. B. Silva, Elson Longo, Ricardo L. Tranquilin, Iêda M. G. Santos, Antônio Gouveia de Souza, Ary da Silva Maia; Synthesis of potassium niobates by microwave assisted solvothermal method

Simposium C - Magnetic Materials
Poster presentation: Aurenice Arruda Dutra Das Merces, Jackeline da Costa Maciel, Luiz Bezerra de Carvalho Júnior; Magnetic particles as affinity matrix for purification of antithrombin

Simposium D - Organic Electronics and hybrids: materials and devices
Oral presentation: Juliana Eccher, Gregorio Couto Faria, Harald Bock, Heinz Von Seggern, Wojciech Zajaczkowski, Wojciech Pisula, Ivan H. Bechtold; Electrical response of a columnar liquid crystal applied in a diode structure
Poster presentation: Francineide Lopes de Araújo, Gustavo Targino Valente, Roberto Mendonça Faria, Francisco Eduardo Gontijo Guimarães; How surface interactions freeze polymer molecules at room temperature: a single molecule approach

Simposium E - Sol-Gel Materials: From Fundamentals to Advanced Applications
Oral presentation: Larissa Brentano Capeletti, Mateus B. Cardoso, João Henrique Zimnoch dos Santos, Wei He; Bioactive hybrid aminopropyl-silica coating to support neuronal growth and suppress astrocyte development

IOP Conf. Series: Materials Science and Engineering **97** (2015) 011001 doi:10.1088/1757-899X/97/1/011001

Simposium F - Anti-fouling Materials and Coatings
Oral presentation: Maria de Fátima B. Souza, Celso Aparecido Bertran, Mônica de Oliveira Penna; Evaluation of polymeric coatings on their efficiency of inhibiting the formation of inorganic scale

***Simposium G - Research Frontiers of Computer Simulations in Materials Science:* Developments and Applications**
Poster presentation: Luis Henrique da Silveira Lacerda, Sergio Ricardo de Lazaro, Renan Augusto Ribeiro; Theoretical investigation of optical and structural properties of Ba-doped ZnO material

Simposium H - Luminescent Materials
Oral presentation: José Miranda Carvalho, Cássio Cardoso Santos Pedroso, Miguel Aguirre Stock Grein Barbará, Pawel Gluchowski, Lucas Carvalho Veloso Rodrigues, Maria Cláudia França da Cunha Felinto, Jorma Hölsä, Hermi Felinto Brito; Luminescence mechanism of titanium doped rare earth oxysulfides phosphors obtained by rapid microwave preparation
Poster presentation: André Lucas Costa, João Paulo Gelamos, Gabriel Mamoru Marques Shinohara, Sergio Antonio Marques Lima, Ana Maria Pires; Amino funcionalization of Y_2O_3:Eu(III) red nanophosphor monitored by luminescence spectroscopy and transmission electron microscopy

Simposium K - Structure-Properties Relationship of Advanced Metallic Materials
Oral presentation: Verona Biancardi Oliveira, Hugo Ricardo Zschommler Sandim; Automatic reconstruction of austenite grain structure in martensitic eurofer-97 steel from electron backscatter diffraction data
Poster presentation: Adriano Gonçalves dos Reis, Danieli Aparecida Pereira Reis, Antonio Jorge Abdalla, Jorge Otubo, Hugo Ricardo Zschommler Sandim; A dilatometric study of the continuous heating transformations in maraging 300 steel

Simposium L - Current Research in Energy Storage Systems
Oral presentation: Sherdil Khan, Marcos Jose Leite Santos, Jairton Dupont, Sérgio Ribeiro Teixeira; Photoelectrochemical study of Ta_3N_5 nanotubes for water splitting
Poster presentation: Suelen Rodrigues Almeida, Carolina Elicker, Bruno Muller Vieira, José Ramon Jurado Egea, Pedro José Sanches Filho, Mário Lúcio Moreira, Sergio da Silva Cava, Cristiane Raubach Ratmann; Hydrogen production from rice husk gray obtained in the pyrolysis process

Simposium M - Nanomaterials for Nanomedicine
Oral presentation: Blanca Azucena Gómez Rodríguez, Manuel Perez Caro, Deborah Zanforlin, Ana Laura Elías, José Luiz Lima, Antonio Gomes Souza Filho, Mauricio Terrones, José Albino Aguiar; Graphene nanoribbons decorated with magnetic Fe_3O_4 nanoparticles for DNA sensing.
Poster presentation: Leonardo Miziara Barboza Ferreira, Natália Noronha Ferreira, Charlene Priscila Kiill, Jovan Duran Alonso, Maria Palmira Daflon Gremião; Adjusting supramolecular electrostatic interactions to produce mucoadhesive nanocarriers for protein delivery

Simposium N - Surface Engineering - functional coatings and modified surfaces
Poster presentation: Rodolfo Bonoto Estevam, Rodolfo Thiago Ferreira, Alan Ben-Hur Bischof, Fábio Santana dos Santos, Cleverson Siqueira Santos, Sérgio Toshio Fujiwara, Karen Wohnrath, Jarem Garcia, Christiana Andrade Pessoa; Preparation and characterization of the LbL films based nanoparticle graphene oxide interacting with 3-n-propylpirydinium silsesquixane chloride

Simposium O - Multifunctional materials derived from clay minerals
Poster presentation: Caroline Silva de Matos, Michele Aparecida Rocha, Christine Taviot Gueho, Fabrice Leroux, Vera Regina Leopoldo Constantino; Layered materials as nanocarriers to bioactive molecules

Simposium P - Advanced Carbon Nanostructures and Composites
Poster presentation: André Luiz Maia Azevedo, Renato Soares de Oliveira, Eduardo Ariel Ponzio, Felipe Silva Semaan; Sensor development exploiting graphite-epoxy composite as electrode material

Simposium S - Advanced Materials
Poster Presentation: Gleison Adriano da Silva, Sérgio Michielon de Souza; Structural, thermal and optical studies of a nanostructured composite formed by binary mixtures of elemental nickel and antimony powders produced by mechanical alloying

The best oral and poster presentations of the whole meeting were chosen among these works, as follows:

Poster:
Simposium P - Sensor development exploiting graphite-epoxy composite as electrode material. André Luiz Maia Azevedo[1], Renato Soares de Oliveira[1], Eduardo Ariel Ponzio[1], Felipe Silva Semaan[1]; [1]Universidade Federal Fluminense

Oral:
Simposium D - Electrical response of a columnar liquid crystal applied in a diode structure; Juliana Eccher[1], Gregorio Couto Faria[2], Harald Bock[3], Heinz Von Seggern[4], Wojciech Zajaczkowski[5], Wojciech Pisula[5], Ivan H. Bechtold[1]; [1]Universidade Federal de Santa Catarina, [2]Universidade de São Paulo, [3]Centre de Recherche Paul Pascal, Cnrs and Univ. Bordeaux, [4]Technische Universität Darmstadt, [5]Max-Planck-Institut For Polymer Research, Mainz

IOP Conf. Series: Materials Science and Engineering **97** (2015) 011002 doi:10.1088/1757-899X/97/1/011002

Peer review statement

All papers published in this volume of *IOP Conference Series: Materials Science and Engineering* have been peer reviewed through processes administered by the proceedings Editors. Reviews were conducted by expert referees to the professional and scientific standards expected of a proceedings journal published by IOP Publishing.

IOP Conf. Series: Materials Science and Engineering **97** (2015) 012001 doi:10.1088/1757-899X/97/1/012001

Synthesis of potassium niobates by the microwave-assisted solvothermal method.

T M Duarte [1], **L M C Honorio** [1], **A S Brito** [1], **J K D Souza** [1], **E Longo** [2], **R L Tranquilin** [3], **A G Souza** [1], **I M G Santos** [1] and **A S Maia** [1,4]
[1] LACOM/INCTMN, Universidade Federal da Paraíba, João Pessoa, PB, Brazil
[2] INCTM/LIEC, Universidade Estadual Paulista, UNESP, Araraquara, SP, Brazil
[3] LIEC/CDMF/INCTMN, Universidade Federal de São Carlos, São Carlos, SP, Brazil
[4] To whom correspondence should be sent. E-mail: arymaia@quimica.ufpb.br

Abstract: Potassium niobates have been synthesized by different methods, but few papers use the microwave-assisted solvothermal method, with possibility of a deeper evaluation in this area. In this work, the synthesis of $KNbO_3$, KNb_3O_8 and $K_4Nb_6O_{17}$ was performed by calcinations of precursors obtained by the microwave-assisted solvothermal method. This study evaluates the influence of various synthesis parameters such as reaction time, pH of the reaction medium, K^+ concentration in the precursor solution. The results show the formation of different niobates, $KNbO_3$, KNb_3O_8, $K_4Nb_6O_{17}$, depending on the short and long-range order of the precursor used in the post annealing process.

1. Introduction

Niobates have wide application in the chemical industry, for presenting versatile structures and thus make possible the development of a wide range of materials with different physicochemical properties [1–3]. For example, $KNbO_3$ exhibits properties and applications as optical devices, acoustic wave surface devices and others [4], while KNb_3O_8 and $K_4Nb_6O_{17}$ exhibit optical and electronic properties, high catalytic activity in hydrogen production by water splitting and are also used in composition of new nanostructured semiconductor materials [5–8].

Different synthesis routes, as solid-state reaction route, sol-gel, hydro/solvothermal, under different conditions are reported for obtaining $KNbO_3$, KNb_3O_8 and $K_4Nb_6O_{17}$, as observed in table 1.

Since the mid 80s, the use of microwave-assisted hydro/solvothermal method has been studied in materials synthesis [9]. This technique has interesting advantages compared to other traditional synthesis processes, especially in relation to a lower energy consumption and time control of the morphology and reproducibility [9–11].

IOP Conf. Series: Materials Science and Engineering **97** (2015) 012001 doi:10.1088/1757-899X/97/1/012001

Table 1. Some methods of synthesis of niobates, reported in the literature.

Material	Methods and References
$KNbO_3$	Solid State [12,13], Sol-Gel [14,15], Conventional Hydro/Solvothermal [4,16], Microwave Assisted Hydrothermal [3,17]
KNb_3O_8	Solid State [18,19], Conventional Hydro/Solvothermal [16,20]
$K_4Nb_6O_{17}$	Solid State [14,21], Sol-Gel [22], Conventional Hydro/Solvothermal [16,23]

2. Experimental

The precursor suspension was obtained using niobic acid and potassium hydroxide in an aqueous/ethanolic medium with a volumetric ratio of 1:1. 70 ml of the suspensions were solvothermalized at a constant reaction temperature of 150 °C. Three parameters were varied during synthesis: reaction time (1, 15, 45 min), K/Nb molar ratio 6.6:1 and 10.5:1) and pH (6, 8 and 14) of the solution, adjusted with acetic acid. Materials were centrifuged and dried at 100 °C for 8 h, and then heat treated at 650 °C for 120 min. The materials were characterized by X-ray diffraction (XRD 6000 – Shimadzu), Raman spectroscopy (micro-Raman LabRAM Horiba Jobin-Yvon, with argon laser) and field-emission scanning electronic microscopy (FEG-VP - Zeiss Supra 35)

3. Results and Discussion

XRD patterns of the material synthesized during 1 to 45 min, at pH = 14 and K:Nb molar ratio of 6.6:1 are shown in figure 1(a). The Lindquist salt ($K_6H_2Nb_6O_{19}.13H_2O$) was obtained in all of the conditions in agreement with the results of Santos et al. [24]. Raman spectrum showed in figure 1b, have good agreement with the results of Farrell et al. [25], as observed in Table 2.

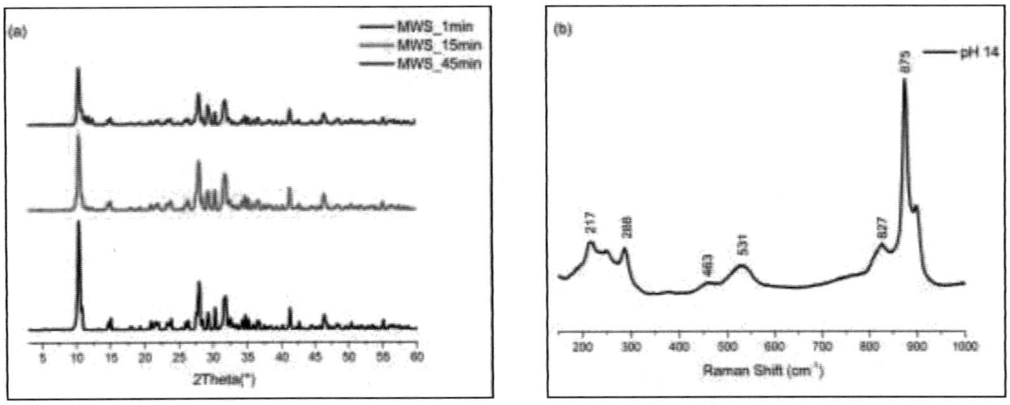

Figure 1. Characterization of the precursor obtained after solvothermalization at pH = 14 and K:Nb molar ratio of 6.6:1. (a) XRD patterns (starred peaks are assigned to the Lindquist salt); (b) Raman spectra.

Table 2. Calculed and observed frequencies (cm^{-1}) for $Nb_6O_{19}^{8-}$.

Modes	Calcd.[a]	Obsd.[a]	Obsd.[b]	Modes	Calcd.[a]	Obsd.[a]	Obsd.[b]
A_{1g}	879	875	875	A_{1g}	500	495	*
E_g	834	823	827	T_{2g}	469	460	463
E_g	726	730	*	A_{1g}	292	290	288
E_g	521	532	531	T_{2g}	224	220	217

* Not cleared observed, [a] From [25], [b] This study.

The post-annealing of the precursor ensures the formation of $KNbO_3$, with orthorhombic structure, in agreement to JCPDS Card 00-032-0822, figure 2(a). The Raman spectra (figure 2.b) is also in excellent agreement with those previously reported in the literature [26,27]. For this authors, the band located at 192 cm^{-1} can be attributed to internal vibrational modes of the NbO_6 octahedron and/or translational modes of K^+ cation, the two bands at 595 and 531 cm^{-1} are two stretching modes (ν_1 and ν_2 respectively), the broad band at 280 cm^{-1} is a bending mode (ν_5) and the weak band at 834 cm^{-1} is a combinational band ($\nu_1 + \nu_5$) [26,27]. FE-SEM micrograph showed a high degree of sintering among particles, as presented in Figure 2c.

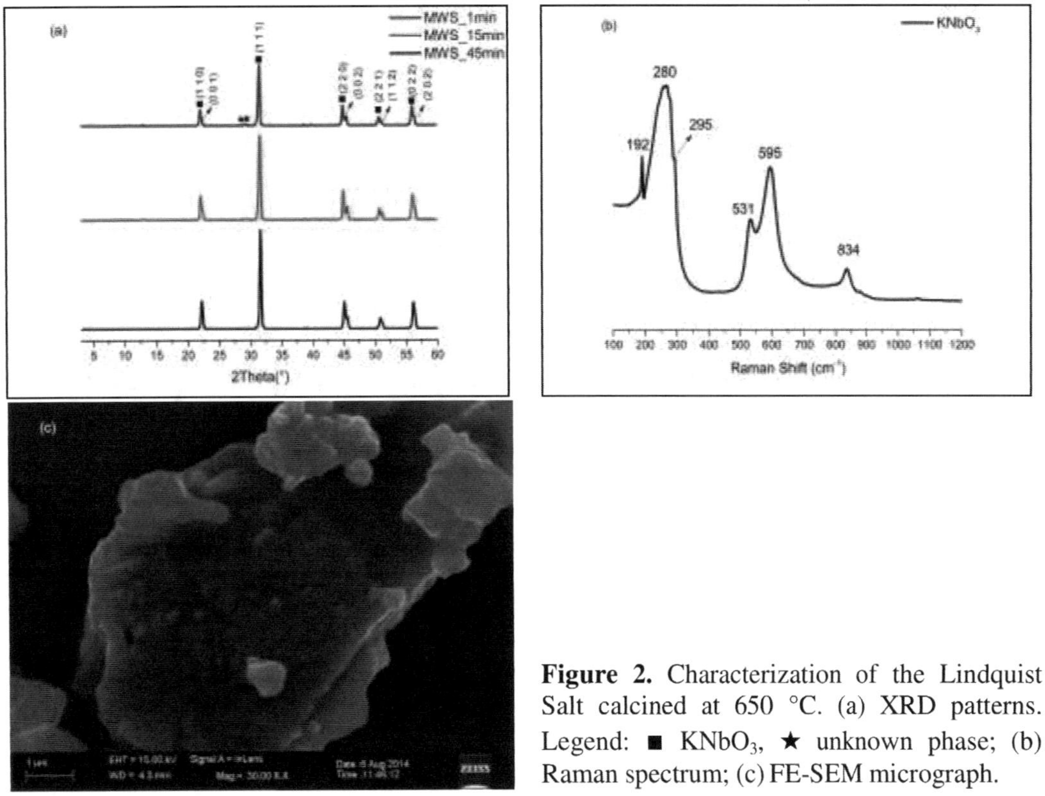

Figure 2. Characterization of the Lindquist Salt calcined at 650 °C. (a) XRD patterns. Legend: ■ $KNbO_3$, ★ unknown phase; (b) Raman spectrum; (c) FE-SEM micrograph.

After modification of the potassium concentration, a K/Nb ratio of 10.5:1 was used with pH values of 6 and 8. The highest amount of potassium in solution associated with the lowest pH led to the formation of an amorphous precursor, ensuring the crystallization of lamellar niobates, KNb_3O_8 and $K_4Nb_6O_{17}$, after post annealing.

Figures 3(a) and 3(b) show the XRD patterns of the niobates, before and after calcination and its indexing according to JCPDS 00-032-0822 ($KNbO_3$), 01-075-2182 (KNb_3O_8), 00-021-1295 ($K_4Nb_6O_{17}$) and 00-021-1295 ($K_4Nb_6O_{17}.3H_2O$)cards. The appearance of the peaks related to the lamellar planes *(0 x 0)*, at angles below 15° is a clear distinction between these XRD patterns and those shown in Figure 2.a. This is the main characteristic of lamellar compounds.

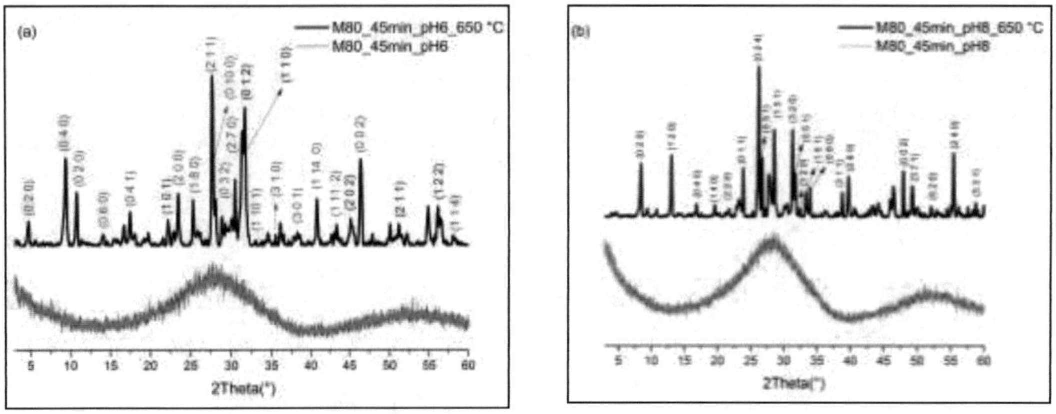

Figure 3. XRD patterns of the precursors obtained from solutions with different values of pH, before and after post annealing at 650 °C. (a) pH = 6; (b) pH = 8. Legend: red - $K_4Nb_6O_{17}.3H_2O$; green - $K_4Nb_6O_{17}$; blue - $KNbO_3$; black – KNb_3O_8 .

It was possible to observe a high dependence on the product direction with the pH since, even when working with higher K/Nb ratio, fact that should favor the formation of $KNbO_3$, by adjusting the solutions pH to less basic values benefited the formation of lamellar species.

Lamellar niobates are characterized by having two different types of octahedra. The first one has shorter Nb-O bonds causing a sharp deformation of such NbO_6 units. These are the octahedra that stuck out into interlayers. The other have longer Nb-O bonds and are slightly distoted [28,29]. These differences are evidenced in the Raman spectra, showed in figure 4. The sharp and intense bands observed at 953 and 881 cm^{-1} (figures 4.a and 4.b respectively) are assigned to the stretching mode of the short Nb-O bonds. The region between 700-500 cm^{-1} is assigned to Nb-O stretching mode of the slightly distorted octahedra. Bands related to the bending modes of Nb-O-Nb linkages are presented in lower Raman shift [28,29].

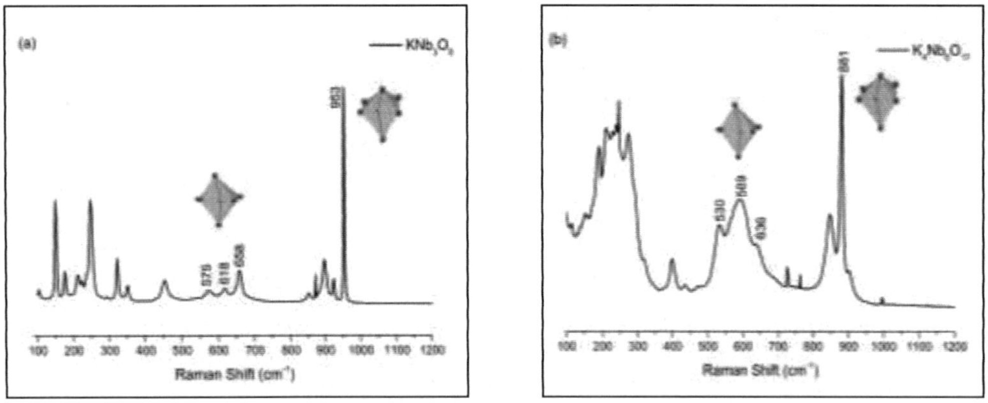

Figure 4. Raman spectra of the lamellar niobates. (a) KNb_3O_8; (b) $K_4Nb_6O_{17}$.

Morphologies of lamellar potassium niobates are presented in the figures 5a and 5b. The formation of superimposed thin small plates is observed as well as some degree of sintering.

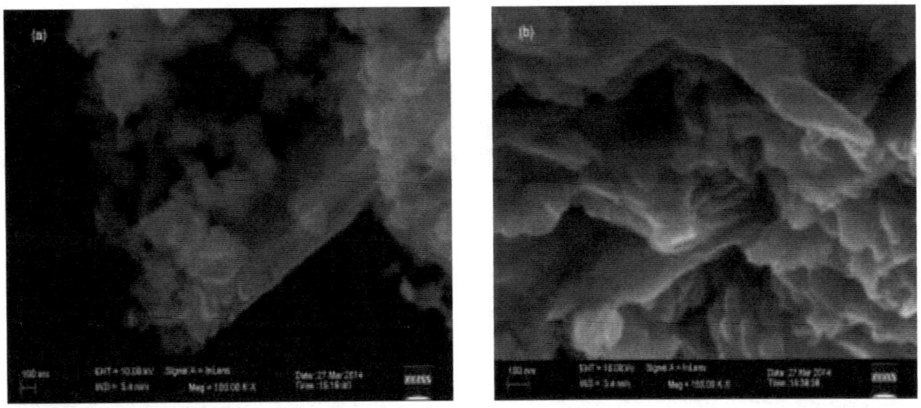

Figure 5. FE-SEM micrographs of the lamellar niobates. (a) KNb_3O_8; (b) $K_4Nb_6O_{17}$.

4. Conclusions

The microwave-assisted solvothermal method was used to obtain precursors for different potassium niobates. Each structure was obtained in a different reaction condition, and the pH of the suspension and potassium concentration determined the niobate structure obtained after post annealing. For instance, post annealing of the highly ordered Lindquist salt formed $KNbO_3$, a material with more ordered octahedra. When a disordered precursor was obtained, post annealing led to lamellar phases which show highly distorted octahedra.

References:

[1] Sugai T and Shindo H 2001 Effects of Adsorbed Water upon Friction at Layered $K_4Nb_6O_{17}.3H_2O$ Surfaces Studied with FFM *Stud. Surf. Sci. Catal.* **132** 897–900

[2] Komarneni S 2003 Nanophase materials by hydrothermal, microwave-hydrothermal and microwave-solvothermal methods *Curr. Sci.* **85** 1729–34

[3] Paula A J, Parra R, Zaghete M a. and Varela J a. 2008 Synthesis of $KNbO_3$ nanostructures by a microwave assisted hydrothermal method *Mater. Lett.* **62** 2581–4

[4] Nakashima K, Fujii I and Wada S 2013 Preparation of $KNbO_3$ nanocubes using a solvothermal method at low temperature *J. Ceram. Soc. Japan* **121** 693–7

[5] Kudo A, Sayama K, Tanaka A, Asakura K, Domen K, Maruya K and Onishi T 1989 Nickel-loaded $K_4Nb_6O_{17}$ photocatalyst in the decomposition of H_2O into H_2 and O_2: Structure and reaction mechanism *J. Catal.* **120** 337–52

[6] Sayama K, Tanaka A, Domen K, Maruya K and Onishi T 1990 Photocatalytic decomposition of water over a Ni-loaded $Rb_4Nb_6O_{17}$ catalyst *J. Catal.* **124** 541–7

[7] Uchida S, Inoue Y, Fujishiro Y and Sato T 1998 Hydrothermal synthesis of $K_4Nb_6O_{17}$ *J. Mater. Sci.* **3** 5125–9

[8] Bizeto M A, Shiguihara A L and Constantino V R L 2009 Layered niobate nanosheets: building blocks for advanced materials assembly *J. Mater. Chem.* **19** 2512

[9] Komarneni S, Noh Y D, Kim J Y, Kim S H and Katsuki H 2010 ChemInform Abstract: Solvothermal/Hydrothermal Synthesis of Metal Oxides and Metal Powders with and Without Microwaves. *ChemInform* **41** 1033–7

[10] Sung H J, Lee J H and Chang J S 2005 Microwave synthesis of a nanoporous hybrid material, chromium trimesate *Bull. Korean Chem. Soc.* **26** 880–1

[11] Prado-Gonjal J, Villafuerte-Castrejón M E, Fuentes L and Morán E 2009 Microwave-hydrothermal synthesis of the multiferroic $BiFeO_3$ *Mater. Res. Bull.* **44** 1734–7

[12] Kinoshita T, Senna M, Doshida Y and Kishi H 2012 Synthesis of size controlled phase pure $KNbO_3$ fine particles via a solid-state route from a core–shell structured precursor *Ceram. Int.* **38** 1897–904

[13] Chaiyo N, Ruangphanit A, Muanghlua R, Niemcharoen S, Boonchom B and Vittayakorn N 2010 Synthesis of potassium niobate ($KNbO_3$) nano-powder by a modified solid-state reaction *J. Mater. Sci.* **46** 1585–90

[14] Amini M M and Mirzaee M 2009 Effect of solvent and temperature on the preparation of potassium niobate by hydrothermal-assisted sol–gel processing *Ceram. Int.* **35** 2367–72

[15] Czechowska K, Psiuk B, Wrzalik R, Szade J, Burdyl M, Śliwa A and Stec K 2014 Preparation of $KNbO_3$ powders by sol-gel method using water-soluble potassium and niobium compounds as precursors *Glas. Phys. Chem.* **40** 88–92

[16] Kong X, Hu D, Wen P, Ishii T, Tanaka Y and Feng Q 2013 Transformation of potassium Lindquist hexaniobate to various potassium niobates: solvothermal synthesis and structural evolution mechanism. *Dalton Trans.* **42** 7699–709

[17] Qin B, Tan G-Q, Miao H-Y, Xia A and Cheng L 2011 Study on Microwave-assisted Hydrothermal Synthesis and the Properties of $KNbO_3$ Powders *J. Inorg. Mater.* **26** 892–6

[18] Bizeto M A, De Faria D L A and Constantino V R L 1999 Organic-inorganic hybrid material synthesized by porphyrin intercalation into a layered niobate host matrix *J. Mater. Sci. Lett.* **18** 643–6

[19] Kudo A and Sakata T 1996 Effect of Ion Exchange on Photoluminescence of Layered Niobates $K_4Nb_6O_{17}$ and KNb_3O_8 *J. Phys. Chem.* **100** 17323–6

[20] Li X, Pan H, Li W and Zhuang Z 2012 Photocatalytic reduction of CO_2 to methane over HNb3O8 nanobelts *Appl. Catal. A Gen.* **413-414** 103–8

[21] Lin H-Y H, Lee T T-H and Sie C C-Y 2008 Photocatalytic hydrogen production with nickel oxide intercalated $K_4Nb_6O_{17}$ under visible light irradiation *Int. J. Hydrogen Energy* **33** 4055–63

[22] Chou H-L, Lee C-C, Chen H-M, Su W-N and Hwang B-J 2011 Characterization of $K_4Nb_6O_{17}$ synthesized by a sol–gel method for H_2 evolution *J. Chinese Inst. Eng.* **34** 3–9

[23] Qu W, Chen F, Zhao B and Zhang J 2010 Preparation and visible light photocatalytic performance of methylene blue intercalated $K_4Nb_6O_{17}$ *J. Phys. Chem. Solids* **71** 35–41

[24] Santos I C M S, Loureiro L ., Silva M F . and Cavaleiro A M V 2002 Studies on the hydrothermal synthesis of niobium oxides *Polyhedron* **21** 2009–15

[25] Farrell F J, Maroni V A and Spiro T G 1969 Vibrational analysis for $Nb_6O_{19}{}^{8-}$ and $Ta_6O_{19}{}^{8-}$ and the the Raman intensity criterion for metal-metal interaction *Inorg. Chem.* **8** 2638–42

[26] Kakimoto K, Akao K, Guo Y and Ohsato H 2005 Raman Scattering Study of Piezoelectric $(Na_{0.5}K_{0.5})NbO_3$-$LiNbO_3$ Ceramics *Jpn. J. Appl. Phys.* **44** 7064–7

[27] Zhang T, Zhao K, Yu J, Jin J, Qi Y, Li H, Hou X and Liu G 2013 Photocatalytic water splitting for hydrogen generation on cubic, orthorhombic, and tetragonal $KNbO_3$ microcubes. *Nanoscale* **5** 8375–83

[28] Li L, Deng J, Yu R, Chen J, Wang X and Xing X 2010 Phase evolution in low-dimensional niobium oxide synthesized by a topochemical method. *Inorg. Chem.* **49** 1397–403

[29] Bizeto M A, Leroux F, Shiguihara A L, Temperini M L A, Sala O and Constantino V R L 2010 Intralamellar structural modifications related to the proton exchanging in $K_4Nb_6O_{17}$ layered phase *J. Phys. Chem. Solids* **71** 560–4

Magnetic particles as affinity matrix for purification of antithrombin

A A D Mercês[1,2], J C Maciel[3], L B Carvalho Júnior[2]

[1]Programa de Pós-Graduação em Biologia Aplicada à Saúde, LIKA/UFPE, Cidade Universitária, 50670-901 Recife, PE, Brazil.
[2]Laboratório de Imunopatologia Keizo Asami, Universidade Federal de Pernambuco, Cidade Universitária, 50670-901 Recife, PE, Brazil.
[3]Universidade Federal de Roraima, Aeroporto, 69310-000 Boa Vista, RR, Brazil.

E-mail: aurenice.dutra@yahoo.com.br

Abstract: Immobilization of biomolecules onto insoluble supports is an important tool for the fabrication of a diverse range of functional materials. It provides advantages: enhanced stability and easy separation. In this work two different magnetic composites were synthesized (MAG-PANI-HS and mDAC-HS) to human antithrombin purification. The magnetic particles (MAG) were obtained by co-precipitation method of iron salts II and III and subsequently coated with polyaniline (MAG-PANI particles). Dacron (polyethylene terephthalate) suffered a hydrazinolysis reaction to obtain a powder (Dacron hydrazide) which was subsequently magnetized (mDAC particles) also by co-precipitation method. Heparan sulfate (HS) was immobilized to MAG-PANI and mDAC retained respectively 35μg and 38.6μg per of support. The magnetic composite containing HS immobilized (MAG-PANI-HS and mDAC-HS) was incubated with human blood plasma (1mL) and then washed with NaCl gradients. Electrophoresis of proteins present in eluates showed bands of antithrombin (58kDa). A reduction in the antithrombin activity was detected in plasma that were incubated in the composites magnetic with HS immobilized, suggesting that the antithrombin was removed of the human blood plasma and then purified. Therefore, the above results suggest that both preparations: MAG-PANI-HS and mDAC-HS are able to affinity purify antithrombin, an important component of blood coagulation.

Keywords: affinity, heparan sulfate, immobilization, magnetic particles, purification.

1. Introduction

Different types of materials have been used as supports for the immobilization of biomolecules, including the supports that exhibit magnetic properties. In addition to being used as adsorbents in magnetic separation, these particles within the micrometer to the nanometer scale, are being used in a growing number of medical applications due to non-toxicity [1,2]. In general, the most suitable inorganic supports have high mechanical strength, good thermal stability, resistance to organic solvents and to attack by microorganisms [3]. Biomolecules immobilized on magnetic supports may be used in the pre analyte concentration, magnetic separation adsorbents, molecular identification of organic and inorganic molecules [4]. Magnetic particles as Dacron ferromagnetic [5] and magnetite coated with polyaniline [6] has been used as supports for immobilization of enzymes.

Biomolecules is immobilized to a solid support by adsorption physical, hydrogen bonding, electrostatic interactions or covalently attached. Covalent immobilization is considered to

result in higher biomolecule activity, reduce nonspecific adsorption, and greater stability of immobilized biomolecules compared to that by physical adsorption, furthermore the biomolecule immobilized covalently cannot be removed from the support by washing with buffers and/or detergents [7].
Affinity separation is a good protein purification method which is based on the formation of reversible and specific complexes between an immobilized molecule and the ligands to be purified. When using a magnetic affinity chromatography has the great advantage of to get a rapid process of separation [8].
Heparan sulfate (HS) is a type of proteoglycan and specific glycoproteins that containing long and sulfated sugar chains that consist of repeating disaccharide units composed of D-glucosamine and D-glucuronic acid or L-iduronic acid. HS have a variety of functions, affecting cell growth, cell adhesion, cell migration, angiogenesis and anticoagulation [9]. So, the HS is able to interact with these and a wide variety of proteins. This work intends to immobilize the heparan sulfate in magnetic particles to affinity purify human blood plasma proteins.

2. Experimental Procedure

2.1. Synthesis of the magnetic particles

The synthesis of Dacron ferromagnetic according to Amaral et al. [5]. First, Dacron was converted to Dacron-hydrazide and then magnetized by co-precipitation method of iron salts II and III. The magnetic particles (MAG) were obtained also by co-precipitation method of iron salts II and III and subsequently coated with polyaniline (MAG-PANI) according to Neri et al [6].

2.2. Immobilization of heparan sulfate on magnetic particles

First, a solution of heparan sulfate 3 mg/mL was functionalized carbodiimide (EDAC) and N-hydroxysuccinimide (NHS). The carboxylic groups of heparan sulfate were activated with EDAC and formed an intermediate compound (O-acylurea) which contains an ester grouping quite reactive and easily undergo hydrolysis. So, was added the NHS that reacted with the ester group of the intermediate compound, leaving it more stable. Thus, the amino group present on the support reacted with the carbonyl of the ester and formed a covalent amide bond [10]. Aliquot 1 ml of heparan sulfate functionalized solution was incubated in each 30 mg of MAG, MAG-PANI and mDAC under light stirring for 72 hours at 25 °C. The amount of heparan sulfate immobilized to the supports was determined according to the methodology used by Oliveira et al. [11].

2.3. Purification of antithrombin

The test of human antithrombin purification by affinity was conducted as follows: for each 30 mg of MAG-PANI-HS and mDAC-HS (magnetic particles with covalently immobilized heparan sulfate) was added 1 mL of human plasma. This mixture was stirred for 60 minutes at 4 °C. Then, with the aid of a magnetic field, the supernatants were collected. Subsequently successive washes with phosphate buffer pH 7.2 in increasing concentrations of NaCl (0.5 M, 1.0 M and 1.5 M) to remove non-adsorbed proteins. A solution of 2.0 M NaCl was added to elute proteins complexed to HS. Electrophoresis of proteins SDS/PAGE (sodium dodecyl sulfate/polyacrylamide gel electrophoresis) 12,5% was performed to the corresponding fractions collected eluates 2M. These human blood plasmas that were incubated in the magnetic derivatives: MAG-PANI-HS and mDAC-HS, were subjected to the antithrombin amidolytic activity kit (TriniCHROM^{TM} Antithrombin IIa - Trinity Biotech SA - New Jersey, USA).

3. Results and Discussion

The particles: Dacron-hydrazide ferromagnetic (mDAC) and magnetite coated with polyaniline (MAG-PANI) had an excellent magnetic activity by their conduct through a magnetic field of 6000 Oe using a magnetic separation system (Abraxis^{TM} LLC), see Fig. 1.

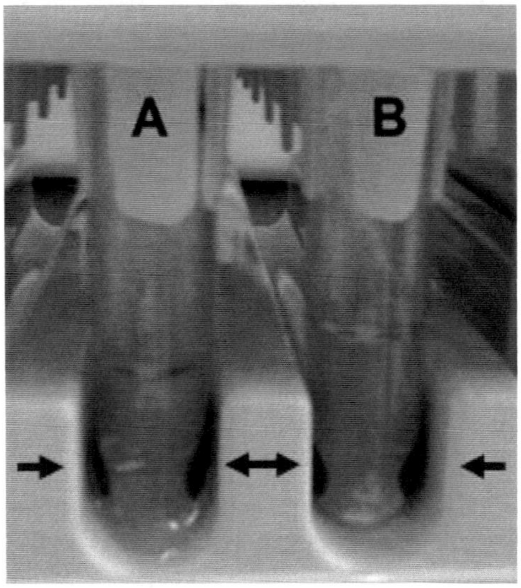

Figure 1: Magnetic separation system with a magnetic field of 6000 Oe (Abraxis^{TM} LLC). The magnetic particles have been attracted by the magnet located on the lateral of the plate (indicated by arrows). A) MAG-PANI; B) mDAC.

Approximately 35μg of heparan sulfate offered was immobilized covalently on MAG-PANI and 38.6 μg in mDAC-HS. An amount of 18 ug of HS was adsorbed in each mg of the MAG particles, this value was lower (compared to MAG-PANI) because MAG particles were not coated with polyaniline. After functionalization with EDAC/NHS [10] the covalent immobilization of heparan sulfate happened by linking its carboxyl groups with the amino groups present in MAG-PANI and mDAC. The electrophoresis of the purified proteins with 2 M NaCl (Fig. 2) in both magnetic supports showed bands corresponding to antithrombin (58kDa) and other bands may be plasma proteins that interact with HS and are identified in subsequent studies. The bands corresponding to the antithrombin were more pronounced in supports containing the covalently immobilized HS: MAG-PANI-HS (Fig. 2A) and mDAC-HS (Fig. 2B). Therefore the other supports did not interact with the antithrombin due to absence or small amount of HS. Heparin and heparan sulfate have structures qualitatively similar [12] and the use of heparin affinity chromatography has been used to purify antithrombin for the past four decades [13]. Here, we study the HS immobilized in magnetic particles to use in the purification process of the AT because was already shown that heparan sulfate also can interact with antithrombin [14].

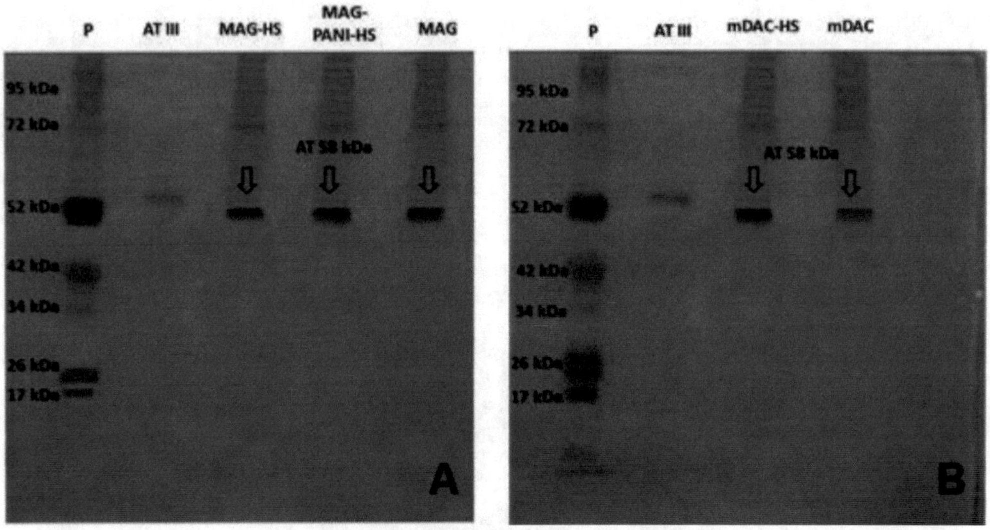

Figure 2: Electrophoresis of proteins obtained in 2M eluates, SDS/PAGE 12.5%. A) MAG-HS (pure magnetite with heparan sulfate adsorbed), MAG-PANI-HS and MAG (pure magnetite). B) mDAC-HS and mDAC (pure ferromagnetic Dacron).

After the purification process, was performed the antithrombin activity of the human blood plasmas that were incubated with MAG-PANI-HS and mDAC-HS. Fig. 3 showed a reduced antithrombin activity of 88% to MAG and 82% to MAG-PANI-HS (Fig. 3A), and 79% in mDAC-HS (Fig. 3B). These reductions suggest that the antithrombin was removed from the human blood plasma and then fixed at HS. Comparing MAG-HS and MAG-PANI-HS, it can be seen that the particle coated with polyaniline showed a more significant AT removal performance, it was probably due to the HS is immobilized covalently on MAG-PANI and only adsorbed (smaller amount) in MAG. So, the magnetic composites with HS have proven useful in the purification of antithrombin process, but according to the results mDAC-HS was a little better because has more HS covalently immobilized and then can removed a large amount of AT when compared to MAG-PANI-HS and MAG-HS.

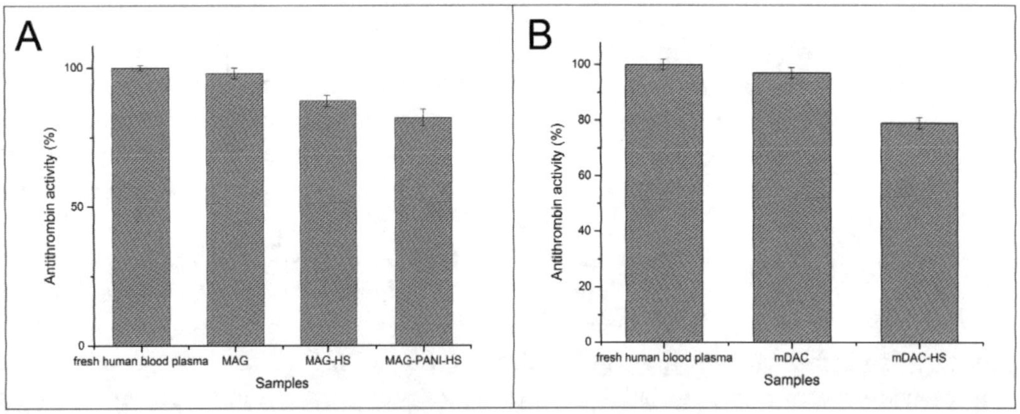

Figure 3: Antithrombin Activity plasma after direct contact with the supports. A) pure magnetite (MAG), pure magnetite with heparan sulfate (MAG-HS) and MAG-PANI-HS. B) pure ferromagnetic Dacron (mDAC) and mDAC-HS.

Angeli et al. [15] demonstrated the use of affinity matrix of ferromagnetic levan composite to purify lectin, this composite used as matrix affinity makes a fast and easy procedure. We obtained similar results, but use magnetic composites for purification of antithrombin human blood plasma. Recently, Lan et al. [8] conducted a study on angiotensin converting enzyme (ACE) purified through magnetic affinity separation, using a magnetic agarose microsphere.
In our study, the affinity matrix is represented by magnetic composite MAG-PANI-HS and mDAC-HS which were capable of interacting with plasma proteins, especially antithrombin. The specificity of these elements led to the removal and identification of specific proteins. Moreover, these magnetic composites described in this work can be applied as affinity matrix to be used in the area of materials science, blood products industry, biochemistry and pharmacology. This is a single and innovative tool and can be exploited in the biotechnology of biomaterials.

4. Conclusions

From the above displayed results one can conclude that the amount of HS immobilized on the particles and the coating of magnetic particles can influence the performance of composites in AT purification process. Thus when comparing the three preparations synthesized in this work, we can see that mDAC-HS proved slightly more efficient because it contained more HS covalently immobilized and so could remove more AT of human blood plasma. Another important fact is that coating of MAG with polyaniline promoted a larger amount of HS immobilized in MAG-PANI particles and thus also more useful in the AT purification process. So, according to the results presented these composites can be applied as magnetic affinity matrices for purification of antithrombin, this human blood plasma protein has an important biological value and the great advantage this method are: easy synthesis and low cost.

Acknowledgments

CAPES, CNPq and FACEPE for financial support.

References

[1] Ito A, Shinkai M, Honda H and Kobayashi T 2005 Medical application of functionalized magnetic nanoparticles *J. Biosci. Bioeng.* **100** 1-11.

[2] Tokoro H, Nakabayashi T, Fujii S, Zhao H and Häfeli U O 2009 Magnetic iron particles with high magnetization useful for immunoassay *J. Magn. Magn. Mater.* **321** 1676-1678.

[3] Bickerstaff Jr G F 1997 Immobilization of Enzymes and Cells *Methods Biotechnol.* **1** 1-11.

[4] Aguilar-Arteaga K, Rodriguez J A and Barrado E 2010 Magnetic solids in analytical chemistry: a review *Anal. Chim. Acta* **674** 157-165.

[5] Amaral I P G, Carneiro da Cunha M G, Carvalho Júnior L B and Bezerra R S 2006 Fish trypsin immobilized on ferromagnetic Dacron *Process Biochem.* **41** 1213-1216.

[6] Neri D F M, Balcao V M, Dourado F O Q, Oliveira J M B, Carvalho-Júnior L B and Teixeira J A 2011 Immobilized β-galactosidase onto magnetic particles coated with polyaniline: Support characterization and galactooligosaccharides production *J. Mol. Catal. B: Enzym.* **70** 74-80.

[7] Tsougeni K, Petrou P S, Awsiuk K, Marzec M M, Ioannidis N, Petrouleas V, Tserepi A, Kakabakos S E and Gogolides E 2015 Direct Covalent Biomolecule Immobilization on Plasma Nanotextured Chemically Stable Substrates *App. Mater. Interfaces* **7** 14670-14681.

[8] Lan X, Liao D, Wu S, Wang F, Sun J and Tong Z 2015 Rapid purification and characterization of angiotensin converting enzyme inhibitory peptides from lizard fish protein hydrolysates with magnetic affinity separation *Food Chem.* **182** 136–142.

[9] Murakami K, Tanaka T, Bando Y and Yoshida S 2015 Nerve injury induces the expression of syndecan-1 heparan sulfate proteoglycan in primary sensory neurons *Neuroscience* **300** 338-350.

[10] Hermanson G T 2008 *Zero Length Crosslinkers* (Bioconjugate Techniques second edition) ed G T Hermanson (New York Academic Press) chapter 3 pp 215-233. Available in http://www.sciencedirect.com/science/article/pii/B9780123705013000035

[11] Oliveira G B, Carvalho Jr L B and Silva M P C 2003 Properties of carbodiimide treated heparin *Biomaterials* **24** 4777–4783.

[12] Brito A S, Cavalcante R S, Palhares L C G F, Hughes A J, Andrade G PV, Yates E A, Nader H B, Lima M A and Chavante S F 2014 A non-hemorrhagic hybrid heparin/heparan sulfate with anticoagulant potential *Carbohydr. Polym.* **99** 372–378.

[13] Miller-Andersson M, Borg H and Andersson L -O 1974 Purification of antithrombin III by affinity chromatography *Thromb. Res.* **5** 439-452.

[14] Chen J and Liu J 2005 Characterization of the structure of antithrombin-binding heparan sulfate generated by heparan sulfate 3-O-sulfotransferase 5 *Biochimica et Biophysica Acta General Subjects* **1725** 190–200.

[15] Angeli R, da Paz N V N, Maciel J C, Araújo F F B, Paiva P M G, Calazans G M T, Valente A P, Almeida F C L, Coelho L C B B, Carvalho Jr L B, Silva M P C and Correia M T S 2009 Ferromagnetic Levan Composite: An Affinity Matrix to Purify Lectin *J. Biomed. Biotechnol.* **2009** 1-6.

IOP Conf. Series: Materials Science and Engineering **97** (2015) 012003 doi:10.1088/1757-899X/97/1/012003

How surface interactions freeze polymer molecules at room temperature: a single molecule approach

F L Araújo[1], G T Valente[1], R M Faria[1] and F E G Guimarães[1]
[1]São Carlos Institute of Physics, University of São Paulo, C.P. 369, *13560-970 São Carlos, SP, Brazil.*

E-mail: franlaraujo@gmail.com

Abstract. The planar conformation assumed by polyfluorenes is used as a model system to investigate how solid surfaces interact with an individual polymer chain as well as to access the effects of substrates and interchain interactions on the polymer conformation. We demonstrate that confocal fluorescence microscopy is very sensitive to characterize photophysical processes associated to conformation assumed by a single conjugated unit of poly (9,9 dioctylfluorene) or PFO. Emissions from interacting molecules have been resolved for the first time between the energies of the planar (β phase) and the non-planar (amorphous) phase that were associated to perturbations of the primary chain conformation. But we have found the planar conformation is assumed by single PFO molecules due to strong interactions with the hydrophilic quartz surfaces. This PFO conformation persists frozen even at room temperature, although emission lines are significantly broadened due to interactions with vibrating substrate atoms.

1. Introduction

The properties of optoelectronic devices based on conjugated polymers show a straight relationship between molecular conformation and their performance. Understanding the nature of the optical processes related to intrinsic molecular arrangement and structure can lead to the development and synthesis of new materials and devices with improved characteristics [1]. Moreover, one of the major challenges in organic electronic is the control of molecular conformation and understand how chain conformation and processes are modified by surfaces and interfaces [2]. In this context, single molecule spectroscopy is a powerful technique in revealing the underlying origin of molecular conformation assumed by the chain at the interfaces and its influence on the photophysical phenomena of conjugated polymers [3,4].

2. Experimental Procedure

Poly (9,9 dioctylfluorene), PFO, was used as a model system deposited on quartz substrate. The polymer solutions were prepared by successive dilution of PFO using the precursor solution with a concentration C_0 = 9.4 μg/ml in chloroform. The polymer samples were deposited on hydrophilic substrate surface by spin-coating technique in static mode. Laser Scanning Confocal Microscopy (LSCM) was used to study the relationship between the photophysical processes and polymeric conformation due interaction with quartz interface.

3. Results and Discussion

Figure 1 shows LSCM images of PFO spincasted on quartz surface from a highly diluted polymeric chloroform solution (9.4 μg/ml). With this procedure we expect to find one PFO molecule on 1.5 μm^2. The Figure 1a was collected with 16 bits and over an area of 2135x135 μm^2 (668x668 pixels), which ensure a continuous intensity scale going in false color from dark read to white with a scanning process that just fills all area without either overlapping or leaving voids between pixels. The density of bright spots in the image is about 1.9 μm^2, close to the expected value for measuring isolated PFO molecules. The digital zoom of Figures 1a (2x2 μm^2 or 10x10 pixels) are depicted in terms of spectral (Figure 1b) and intensity images (Figure 1c). The high dispersion in spectral shapes and positions and the high emission efficiency, both evaluated for each bright spot, are consistent with typical single molecule emission characteristics.

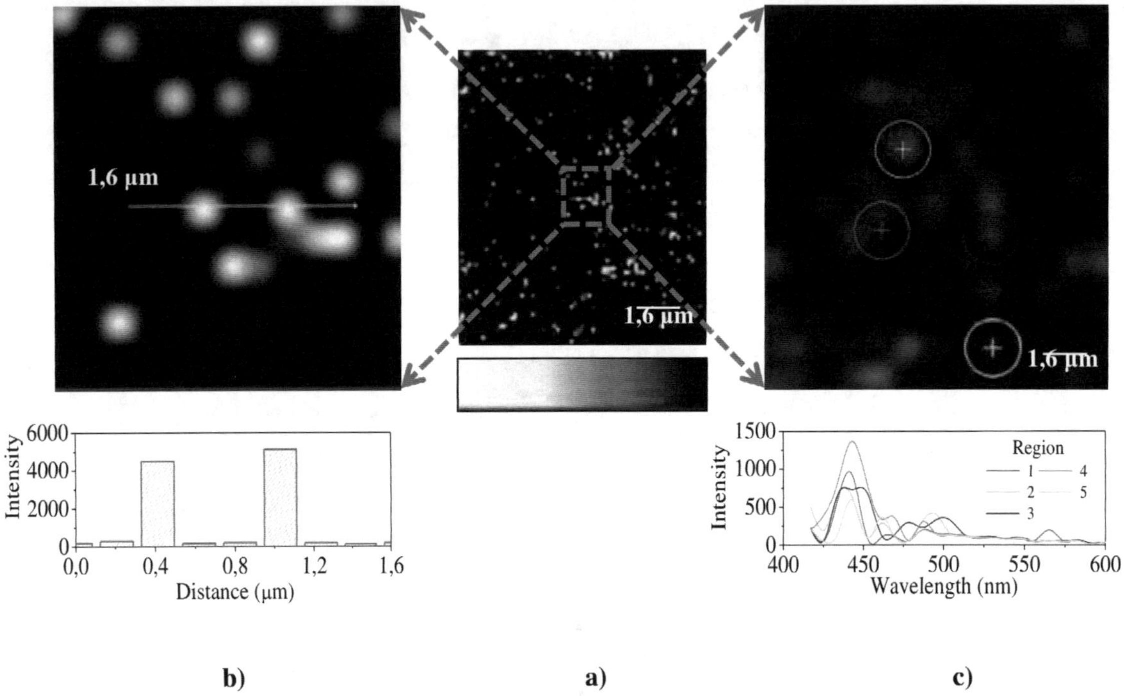

Figure 1: a) LSCM image of PFO spincasted on quartz surface from a highly diluted polymeric chloroform solution. b) Digital zoom (2x2 μm^2 or 10x10 pixels) depicts details in terms of spectral and c) intensity images. Corresponding spectral data for different pixels (colored circles) and intensity profile over a 1.6 μm line are also shown.

The emission characteristics of PFO single molecules can be seen in greater detail in Figure 2. 10 PFO spectra recorded from isolated pixels on Figure 2a are compared in Figure 2b giving evidence the zero-phonon emissions in the range of the expected energies for the planar (β phase around 2.81eV) and non-planar (disordered around 2.94 eV) conformations [5]. It is interesting to see the variety of spectral shapes and positions and that the planar conformation sets the lowest energy for the PFO emissions. The emission lines detected bellow 2.81 eV correspond to phonon replicas whose intensity demonstrate very low phonon coupling to PFO electronic states due to strong steric constraints between isolated molecules and the substrate. Zero-phonon lines positioned between the β and disordered phases can be associated to perturbations of the planar conformation associated to interchain interactions or partial loss of steric constraints with substrate.

Figure 2c compares PFO spectra of a 40 nm thick film (grey line) having only the non-planar phase, of a 1 nm thick film (orange line) and of a single molecule (green line) measured at 300 K with the spectrum of the same 1 nm thick film measured at 5 K. PFO spectrum having non-planar phase are broad and displaced to the high energies and present intense vibronic features at 300 K due to strong electron-phonon coupling. The 1 nm thick ultrathin film shows well resolved vibronic structures and the zero-phonon line positioned between the planar and non-planar phases. This intermediate emission feature can be associated to perturbed planar phase as a consequence of interchain interaction. However, the single PFO molecule present thinner 300 K emission line positioned at same position as the emissions of PFO ultrathin film measured at 5K. This find indicates that PFO single molecules preserve the planar phase even at 300 K due to strong interaction with hydrophilic quartz surface. In addition, the broadening of the single molecule zero-phonon line can be accounted to the perturbation introduced by the vibrational motion of the quartz atoms. The additional thermal energy at room temperature is not enough to break the PFO planar conformation. This means that PFO molecule is totally rigid (frozen) on the substrate surface at room temperature.

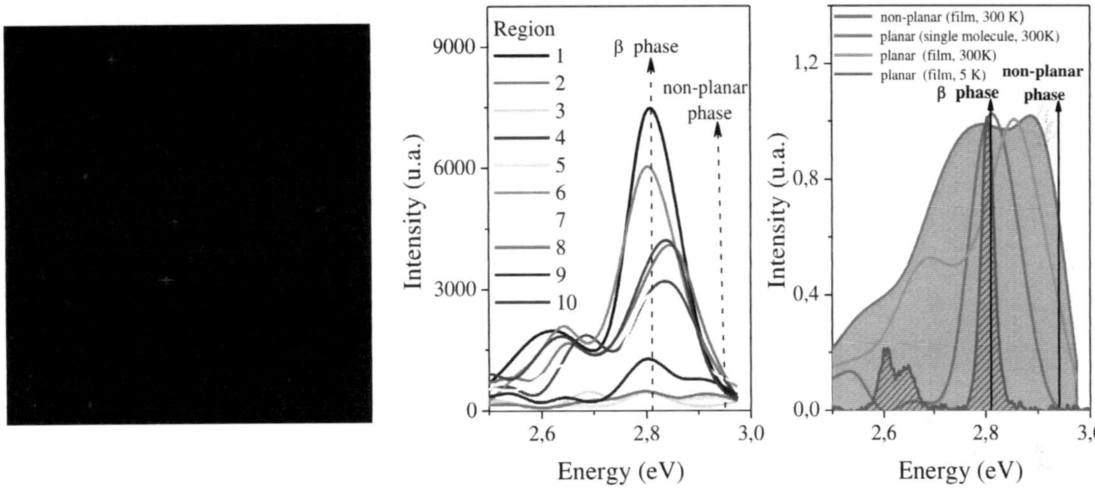

a) **b)** **c)**

Figure 2: a) LSCM spectral image of isolated PFO molecules. b) 10 PFO spectra recorded from isolated pixels on this confocal image. c) PFO spectra of a 40 nm thick film (grey line) having only the non-planar phase, of a 1 nm thick film (orange line) and of a single molecule (green line) measured at 300 K with the spectrum of the same 1 nm thick film measured at 5 K.

4. Conclusions

In this study we used Laser Scanning Confocal Microscopy (LSCM) to investigate the influence of interaction between solid surfaces with a polymer chain and how this affects the polymeric conformation at interface. The microscopy results described here shown that LSCM technique is sufficient sensitivity to measure isolated molecule. Moreover, we find that the substrate imposes a form of freeze conformational in isolated PFO molecule at room temperature. Thermal perturbation due to quartz atoms is not enough to break the π conjugation (planar structure) at 300 K.

5. References

[1] Ariu M, Lidzey D G, Sims M, Cadby A J, Lane P A and Bradley D D C 2002 *J. Phys.: Condens. Matter* **14** 9975.
[2] Vogelsang J and Lupton J M 2012 *J. Phys. Chem. Lett.* **3** 1503.
[3] Barbara P F et al. 2005 *Acc. Chem. Res.* **38** 602.
[4] Lupton J M 2010 *Adv. Mater.* **22** 1689.
[5] Ariu M, Sims M, Rahn M D, Hill J, Fox A M, Lidzey D G 2003 *Physical Review B* **67** 195333-1.

Acknowledgments

This work was supported by CNPq, CAPES, FAPESP and INEO.

IOP Conf. Series: Materials Science and Engineering **97** (2015) 012004 doi:10.1088/1757-899X/97/1/012004

Evaluation of polymeric coatings on their efficiency of inhibiting the adhesion of inorganic scale

M F B Sousa[1,3], C A Bertran[1], M O Penna[2]

[1]Universidade Estadual de Campinas, Instituto de Química, Campinas, SP, Brazil
[2]CENPES – Petrobras, Rio de Janeiro, RJ, Brazil
[3]to whom any correspondence should be addressed: mfabr@iqm.unicamp.br

Abstract: The use of polymer coatings allows modulating wettability and can contribute decisively to mitigate the formation of scale in pipes and equipment. In this work critical surface energy (CSE) and surface free energy (SFE) were determined for several polymer coatings (epoxy phenol novolac resins and fluoropolymers), aiming the evaluation of their anti-fouling capacity against brines reproducing formation waters from pre salt wells. Coatings with lower values for SFE and CSE presented the lowest degree of fouling as determined by wet analysis using inductively coupled plasma atomic emission spectroscopy (ICP - AES).

1. Introduction

Inorganic scale on the surface of equipment used in oil extraction originates from crystallization reactions involving ions present in the formation water. Scale deposition and adhesion are consequences of the properties of the interface between the solid surface and the scaling brine [1,6].

One of the interface features that contribute to the adhesion is wettability [7]. This property is determined by measuring contact angle, which is the angle that a drop of a probe liquid forms when in contact with a solid surface. Two quantities can be calculated from wetting data: the surface free energy (SFE) and the critical surface energy (CSE). The critical surface energy also known as the critical surface tension, characterizes the wettability of a surface by several liquids. Solids presenting high CSE are wettable by most liquids, while solids having low CSE are not wettable by most liquids. In turn, the surface free energy (SFE) is defined as the excess energy at the surface of a material compared to its bulk.

The use of polymer coatings allows modulating wettability and can contribute to mitigate the formation of scale in pipes and equipment, however the literature reports only few studies dealing with the application of polymeric coatings as anti-fouling surfaces for use in the oil and gas industry. In this work the critical surface energy and surface free energy were determined for seven polymer coatings from four different suppliers aiming the evaluation of their anti-fouling capacity when exposed to brines reproducing formation waters from pre salt wells in Brazil. Five epoxy phenol novolac resins (EPNA and EPNB from supplier 1; EPNC and EPND from supplier 2; EPNE from supplier 3) and two fluoropolymer resins (FPA from supplier 4 and FPB from supplier 2) were investigated. Stainless steel (SS) was used as reference.

2. Experimental

2.1. Contact angle measurements

Contact angle measurements were performed on the flat base of cylindrical coupons or on flat plates applying the sessile drop technique using a tensiometer/goniometry Attention Theta (BiolinScientific). Samples were rapidly cleaned with ethanol, rinsed with deionized water and ultrasonicated for 5 min in deionized water, then the contact angles were measured for three probe liquids. The used liquids and their average volumes were: water (5.5 μL), glycerol (4.4 μL) and ethylene glycol (7.0 μL). Hamilton syringe with stainless steel needle (out diameter = 0.72 mm; inner diameter = 0.21 mm) was used to dispense the drops.

2.2. Tests for the formation of scale deposits

Separated synthetic cations- and anions-containing brines, simulating composition of formation water (FW) with high sulfate and strotium content, were mixed and used for conducting scaling experiments (figure 1). Cations- and anions- containing solutions were prepared using analytical grade reagents dissolved in deionized water, and were filtered by membranes with pore size of 0.45 micrometers. Equal volumes (50.00 mL) of each of these two solutions were mixed and only the bottom of the coupon, connected to a rotor from PINE was exposed to the resulting mixture for 24 hours at 56 °C. A rotational speed of 200 rpm was applied. After being removed from the scaling medium, the coupon was slowly and carefully washed 4 times by immersion into deionized water. Then it was kept in an oven at 60 °C during 4 hours before preparation for ICP analysis.

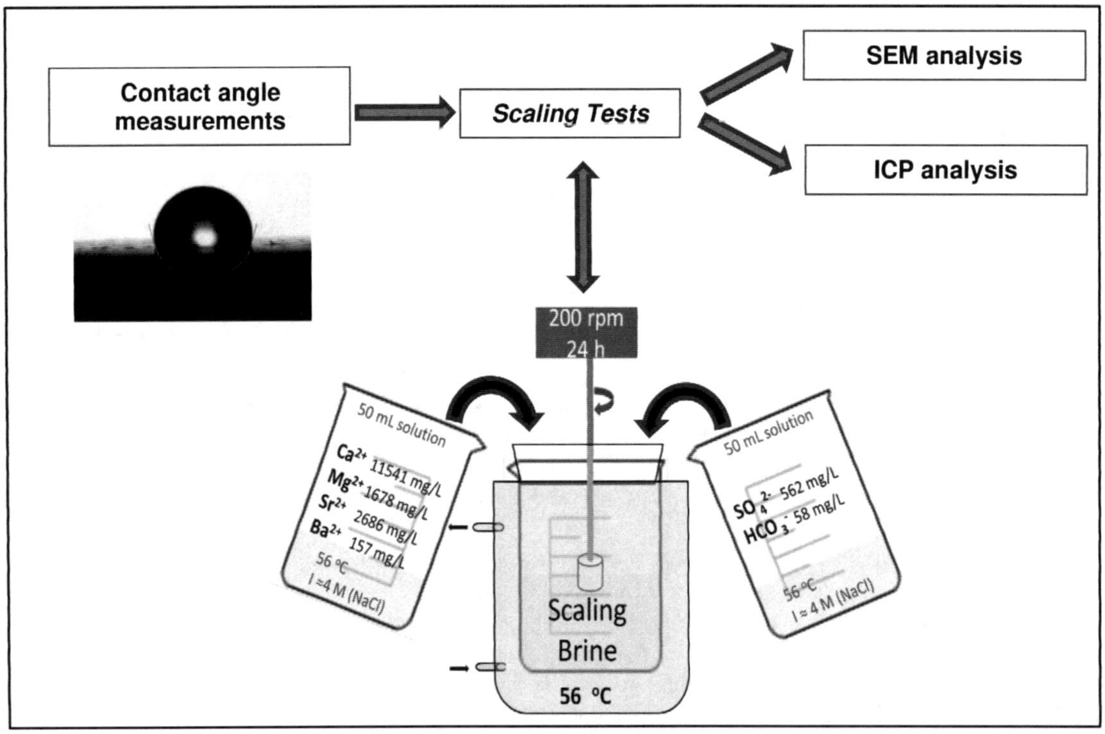

Figure 1. Overall experimental set-up.

2.3. Analysis by Scanning Electron Microscopy (SEM)

After the scaling experiments the coupons with scale deposits were washed with deionized water, dried at 60 °C and examined in a scanning electron microscope (JEOL SM-6360LV - JEOL Ltd.). Samples were imaged by SEM secondary electrons applying an accelerating voltage of 2 kV or 5 kV.

3. Results and Discussion

For the estimation of critical surface energies (CSEs), Zisman plot – type graph (figure 2) were done by plotting liquid surface tension (γ) on the X-axis, and the cosine of the corresponding contact angle on the Y-axis [8]. The extrapolation of the cos θ *vs* γ curve to cos θ = 1 (corresponding to the zero contact angle, or the complete wetting limit) gives the value of the CSE of the tested surface. Three pure probe liquids were used with surface tensions ranging from 72,8 mN m^{-1} to 48,0 mN m^{-1} (water, glycerol and ethylene glycol), which allowed a good precision in the extrapolation process as described above. The surface tension values used were: Υ_{water} = 72.8 mN m^{-1}; $\Upsilon_{glycerol}$ = 63.1 mN m^{-1}; $\Upsilon_{ethylene\ glycol}$ = 48.0 mN m^{-1}. It is necessary to point out that the critical surface energy (CSE) is in fact a wettability range showing the maximum surface tension a liquid may have to ensure complete wetting of a given material.

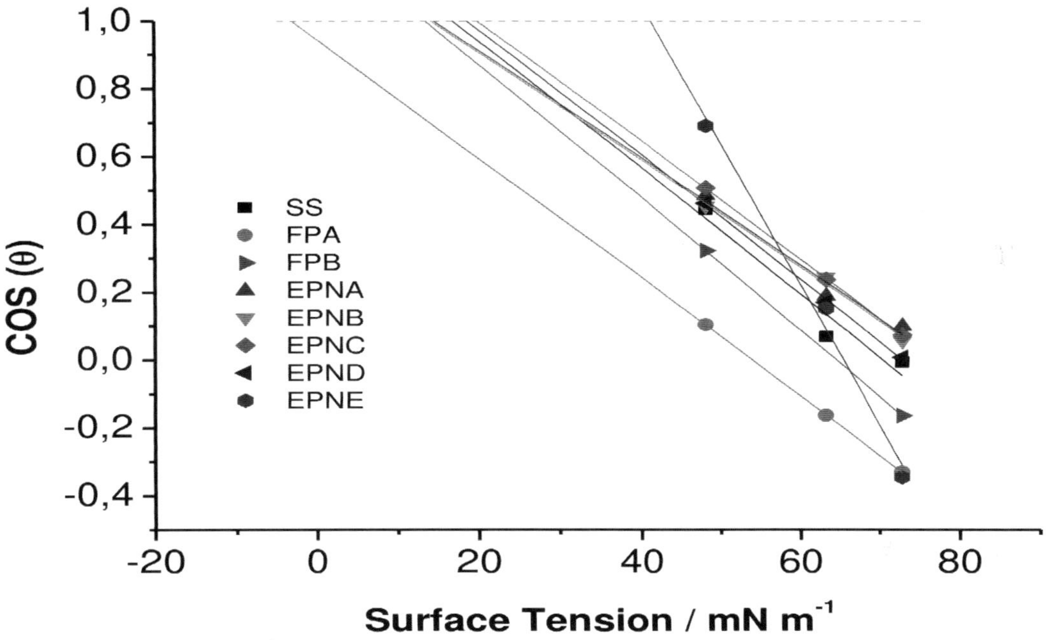

Figure 2. Zisman plot – type graph: extrapolation of the cos θ *vs* γ curve to cos θ = 1 (corresponding to the zero contact angle, or the complete wetting limit for each coating).

It can be observed in Figure 3 that the fluor-based coatings, FPA and FPB, are hydrophobic, i.e., each of them presents water-contact angle higer than 90°, this behavior is expected for fluoropolymers coatings [9], but their ECS and SFE values are very similar those presented by epoxy phenol novolac resins EPNA and EPNB. All other coatings form one block with exception of EPNE, which shows a peculiar behavior, i.e. it presented high CSE, but has good discrimination for the probe liquids with hidrophobicity similar to FPA coating

Although in theory only two liquids would be enough for the determination of CSEs, in practice, this is not true; if only few different liquids are used, incorrect results and even negative values can be obtained, as it occurred to FPA. For this reason, another approach to characterize these surfaces considering their surface energy was also applied.

Many theories using different approaches are used for these calculations [10-12] involving testing the solid against two liquids whose polar and dispersive components of their surface tension are known. In this work the surface free energy (SFE) was calculated by using the OWRK/Fowkes equation (1). In this case, the adhesion energy is a sum of two partial contributions, each of which is expressed as the geometric mean of two adhesion parameters, γ_L and γ_S, characterizing the powers of adhesion of the liquid and of the solid, respectively. The superscripts "*d*" and "*p*" refer to the dispersive and polar components of the probe liquids (Table 1) and of the solid surfaces (to be calculated). Results for this later methodology are presented in figures 3 and 4.

$$\gamma_L(1 + cos\theta) = 2\left[(\gamma_L^d \gamma_S^d)^{\frac{1}{2}} + (\gamma_L^p \gamma_S^p)^{\frac{1}{2}} \right] \quad (1)$$

Table 1 - Surface tension parameters for water and ethylene glycol [4]

Probe liquid	γ_L (mN/m^{-1})	γ_L^d (mN/m^{-1})	γ_L^p (mN/m^{-1})
Water	72.8	21.8	51.0
Ethylene glycol	48.0	29.0	19.0

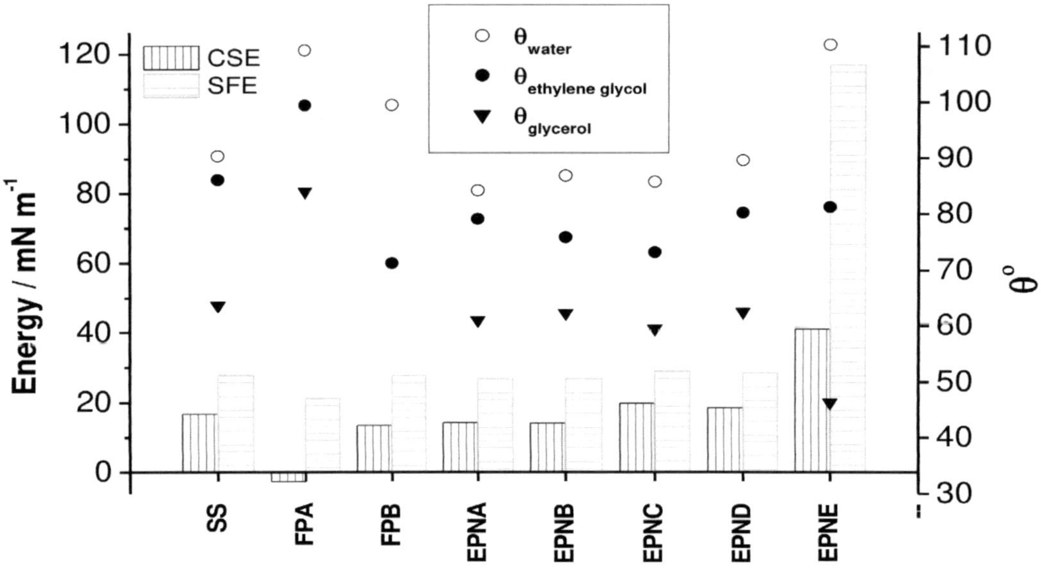

Figure 3. Values of contact angles, Critical Surface Energy (CSE) and Surface Free Energy (SFE) for each coating and the reference SS.

In figure 4 SFE for each coating is shown divided into its two components (polar and dispersive), as well as the content of Ca, Sr and Ba present in the deposits formed onto these coatings after their contact with the synthetic formation water for 24 hours. Coatings FPA, FPB and EPNB achieved higher inhibition efficiency, and EPNE was the coating with the highest degree of fouling.

Coatings presenting lower values for the polar component generally showed the lowest degree of scale compared to stainless steel as can be seen by ICP results (figure 4) and also in the micrographs presented in figure 5. In general, the deposit formed is not uniformly distributed over the surface of the coupon, therefore one should keep in mind that the analysis by ICP takes into account the total deposit while a SEM image is more sectorial and, depending on the magnification, shows only a small part of the whole.

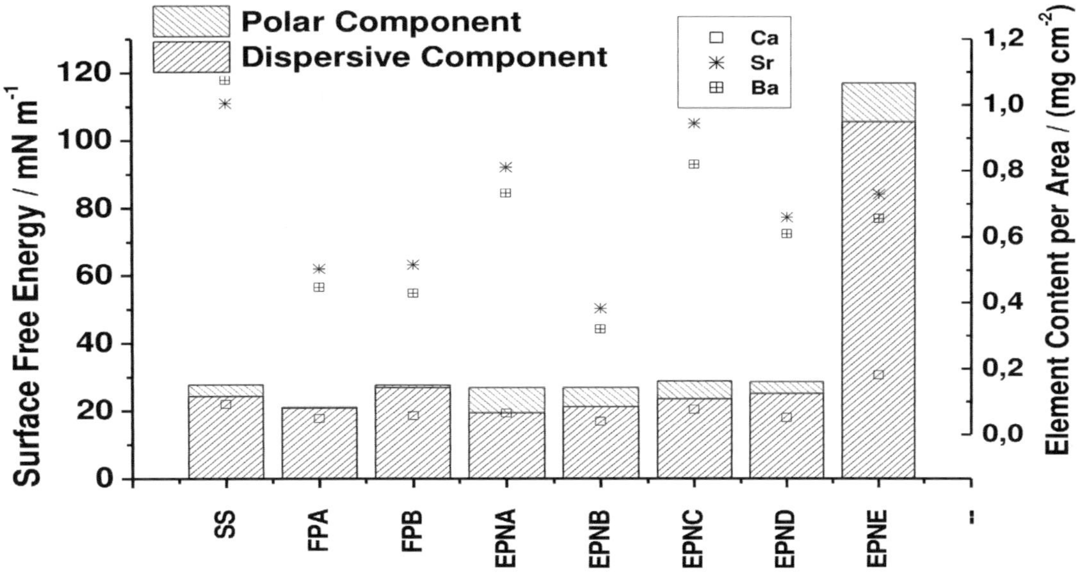

Figure 4. Surface free energy and its components. Elemental composition of scale deposit per coating area, for each coating

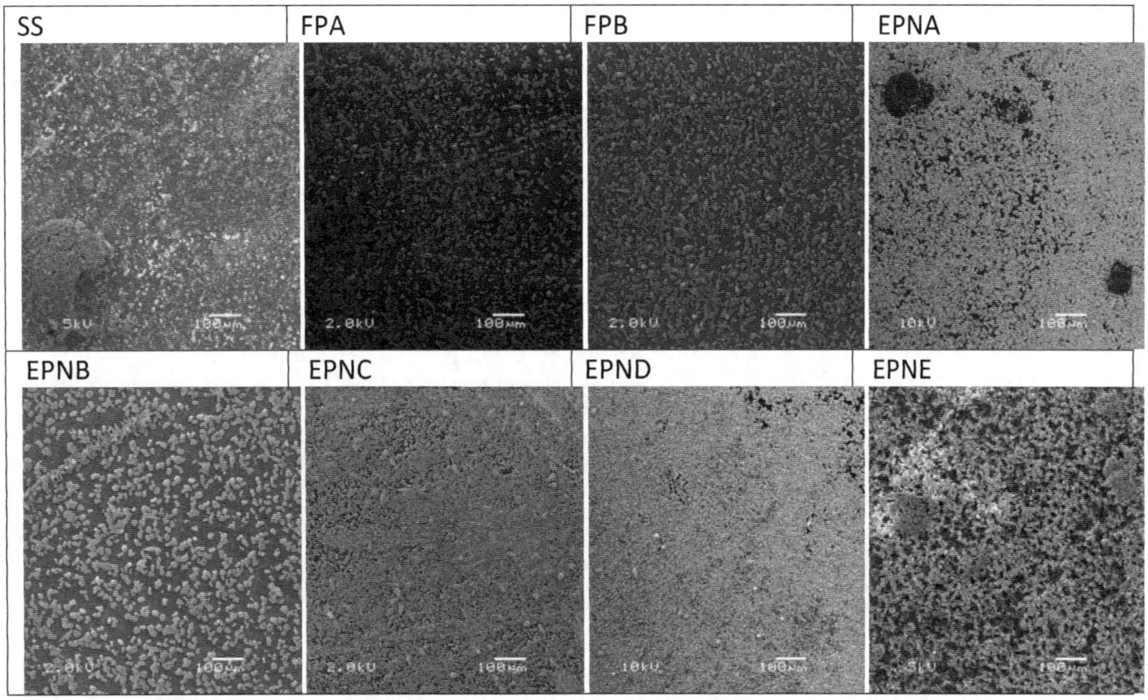

Figure 5. SEM images showing scale deposited onto bare SS and coated coupons.

IOP Conf. Series: Materials Science and Engineering **97** (2015) 012004 doi:10.1088/1757-899X/97/1/012004

4. Conclusion

Taking into account that all specimens were exposed to the formation water under the same conditions of temperature, stirring and time, it can be concluded that in relation to the deposition of calcium carbonate only SS presents some selectivity. However for barium and strontium sulfates there is a significant variation in the degree of deposition onto each of the evaluated coatings. Fluor-based coatings FPA and FPB presented lowest values the polar component in SFE. In general, considering the results of ICP and SEM, there are two groups of coatings. For the same manufacturer, coatings with lower SFE polar component presented less scale deposition. Coating with microscopically rougher surface like EPNE presented higher scale deposition.

Acknowledgments

The authors wish to acknowledge Petrobras-CENPES and ANP for their funding (Grant No. 0050.0077857.12.9)

References

[1] Gunn D J 1980 Effect of surface roughness on the nucleation and growth of calcium sulphate on metal surfaces *J. Cryst. Growth* **50** 533-7.

[2] Lin N H, Shih W-Y, Lyster E and Cohen Y 2011 Crystallization of calcium sulfate on polymeric surfaces *J. Colloid Interf. Sci.* **356** 790-7.

[3] Al-Anezi K, Johnson D J and Hilal N 2008 An atomic force microscope study of calcium carbonate adhesion to desalination process equipment: effect of anti-scale agent *Desalination* **220** 359-70.

[4] Cai Y, Liu M and Hui L 2014 Observations and mechanism of CaSO4 fouling on hydrophobic surfaces *Ind. Eng. Chem. Res.* **53** 3509-27.

[5] Zisman W A 1963 Influence of constitution on adhesion *Ind. Eng. Chem.* **55** 18-38.

[6] Kazi S N, Duffy G G and Chen X D 2010 Mineral scale formation and mitigation on metals and a polymeric heat exchanger surface *Appl. Therm. Eng.* **30** 2236-42.

[7] Szymczyk K, Zdziennicka A, Krawczyk J and Jańczuk B 2012 Wettability, adhesion, adsorption and interface tension in the polymer/surfactant aqueous solution system. I. Critical surface tension of polymer wetting and its surface tension *Colloid Surface* A **402** 132-8.

[8] Kabza K, Gestwicki J E and McGrath J L 2000 Contact angle goniometry as a tool for surface tension measurements of solids using Zisman plot method J. Chem. Edu. 77 63-5.

[9] Chaudhuri R G and Paria S 2009 Dynamic contact angles on PTFE surface by aqueous surfactant solution in the absence and presence of electrolytes *J. Colloid Interf. Sci.* **337** 555-62.

[10] Kwok D Y, Lam C N C, Li A, Leung A , Wu R, Mok E and Neumann A W 1998 Measuring and interpreting contact angles: a complex issue *Colloid Surface* A **142** 219-35.

[11] Siboni S, Della Volpe C, Maniglio D and Brugnara M 2004 The solid surface free energy calculation II *J. Colloid Interf. Sci.* **271** 454-72.

[12] Della Volpe C, Maniglio D, Brugnara M, Siboni S and Morra M 2004 The solid surface free energy calculation I. In defense of the multicomponent approach *J. Colloid Interf. Sci.* **271** 434-53.

Theoretical investigation of optical and structural properties of Ba-doped ZnO material

L H S Lacerda[1,2], S R de Lazaro[1], R A P Ribeiro[1].
[1]*Ponta Grossa State University, Department of Chemistry, Ponta Grossa, PR, Brazil.*
[2] To whom correspondence should be sent. E-mail: srlazaro@uepg.br

Abstract. The doping process is a technique widely used for improving the properties of semiconductors. Through insertion of doping controlled amount is possible change drastically the electronic, optical and structural properties of a material. This work focuses on effects of Ba atoms insertion on wurtzite-ZnO structure at 12.5% amount. The results showed that the presence of Ba in low quantity cause increase in the lattice parameters and decrease in band-gap in relation to the ZnO material. In the percentage of 12.5 %, the doping is noted as a potential alternative for application in opt-electronic devices, electronic devices, solar cells and photocatalytic process.

1. Introduction

The doping process consists in adding a controlled amount of impurities in semiconductor structure. This process changes drastically the properties of a semiconductor material without changing its crystalline structure. Through integrated impurities in the semiconductor structure is possible to control their properties. For this reason, the semiconductor materials can be employed at several electronic devices [1, 2]. Among the semiconductors materials, the wurtzite ZnO has attracted interest in recent years because electronic, optical and piezoelectric properties. The ZnO is a n-type semiconductor of direct band-gap of 3.37 eV [3-5], which presents electronic, optical, ferro- and piezoelectric single properties [3, 6-8]. Due to these properties, such material is employed on development of several electronic devices [4, 7-16]. In this crystalline structure, the Zn and O atoms are observed in a hexagonal-close-packed lattice and both atoms occupy the center of tetrahedral site; hence, the wurtzite-ZnO is relatively open, once all octahedral and half tetrahedral sites are empty. Thus, the doping process can be made without the need of complex experimental procedures [6, 10, 11, 17-24]. The incorporation of external dopants improves the materials properties according to impurity type, for instance, group I and II elements causes the improvement of ferroelectric properties and enables a band-gap engineering.[25]; while the group III elements provides transparent-conducting-oxide property [10] and transition metals make of ZnO a multiferroic semiconductor.

Nowadays, the ZnO material doped with several elements inserted in the crystalline structure, but there is very low manuscripts focused on Ba-doped materials. In 2013, Haja Hameed and coauthors [26] studies the doping of ZnO materials with alkaline metals and observed the band-gap decrease. At same way, Srinet and coauthors [25] synthetized the Ba-doped ZnO at 5 %. Their results indicate the improvement of electronic, optical and ferroelectric properties of Ba-doped sample regarding to ZnO materials. Others papers showed results for ZnO/$BaTiO_3$ heterojunction [27-30] and BaO/ZnO interfaces [31].

However, DFT investigations about Ba-doped ZnO systems were not found at present moment. Theoretical methods based on quantum mechanical simulations are an important tool to evaluate material properties, mainly at the molecular level. Studies based on these methods are performance to evaluate the electronic, optical, structural, pyro-, piezo- and ferroelectric properties. Such properties are intrinsically dependent on composition, atomic organization and chemical bonds, once these factors change the charge density on materials. Therefore, we propose theoretical calculations employing the Density Functional Theory (DFT) allied to hybrid functional B3LYP to investigate the effects of barium-doping of 12.5% on electronic, optical and structural properties of ZnO materials.

2. Computational Methodology

Theoretical investigation of 12.5% Ba-doped ZnO (ZnO:Ba) was performed from periodic models analysis. The simulations were based on Density Functional Theory (DFT) allied to B3LYP [32, 33] parametrized hybrid functional, both employed in CRYSTAL09 [34, 35] software package. The model for ZnO was based on wurtzite structure regarding to literature parameters [8]; while the Ba-doped model was developed by unit cell expansion in the direction of a, b and c axes resulting in a 2x2x2 supercell. One Zn atom was replaced by one Ba atom in the supercell resulting in a concentration of 12.5%. In all models, the Zn and O atoms were described by TZVP [36] all electron basis set; whereas, the Ba atoms were described by HAYWSC [37] pseudopotential employed in Zagorac basis set [38].

The calculation parameters applied were: the SCF convergence factor was truncated in 10^{-8} Hartree; Mohnkhost-Pach method [39] was defined as eight; whereas, total and partial projected Density of States (DOS) were performed using last five energy levels localized on top of valence band (VB) and the first five bands from bottom of conduction band (CB). Such calculation parameters of computational simulation were performed in CRYSTAL09 software package, the structural analysis was made from XCrysden software package [40, 41]. Both software were used on Ubuntu Linux in computational cluster composed by seven quad-core AMD computers with 32 GB of RAM and data storage capability of 3.5 TB.

3. Results and Discussion

The presence of the Ba atoms in the ZnO crystalline structure is responsible for increasing the lattice parameters (Table 1); such increase for these values is dependent on the largest bond length of Ba – O bonds (2.462 Å) in relation to Zn – O bonds (1.980 Å) and also causes slight variations for unit cell angles (α, β and γ). Such changes show that Ba-doping at 12.5% is possible in wurtzite structure because lattice parameters and cell angles were low changed. Another result that help to indicate a possible phase transition is the tetragonality factor (c/a), which is obtained through ratio between a and c lattice parameters of the primitive cell. In Table 1 is showed a slightly change in the c/a factor indicating that there was no a phase transition.

Table 1. Theoretical results for lattice parameters (in Å) and structural angles (in degrees) for ZnO and ZnO:Ba materials.

Model	a	b	c	α	β	γ	c/a
Exp.[a]	3.249	3.249	5.204	90	90	120	1.602
ZnO	3.266	3.266	5.279	90	90	120	1.616
ZnO:Ba	3.364	3.399	5.464	90	90.060	120.346	1.624

[a] Özgür *et al.*, 2005

The Figure 1 shows total and partial projected DOS analysis for ZnO and ZnO:Ba materials. For ZnO material (Figure 1a), the atomic contribution on top of VB for ZnO was projected between -10 and -5 eV (last five energy levels) been composed by O 2p orbitals and Zn 3d orbitals; whereas, the bottom of CB was described from -2 to 12 eV with contributions of O 2p, 3s orbitals and Zn 4s, 4p orbitals. From doping process, the top of VB was displaced between -6 and -5 eV been composed by O 2p orbitals and Zn 3d orbitals and Ba 4d orbitals; whereas, the bottom of CB was calculated from -2 to 2 eV with contributions O 2p, 3s orbitals and Zn 4s, 4p orbitals. Ba atoms have contributed in VB through 4d orbitals; whereas, 5sp and 6sp hybrid orbitals contribute to CB at -2 to 2 eV range.

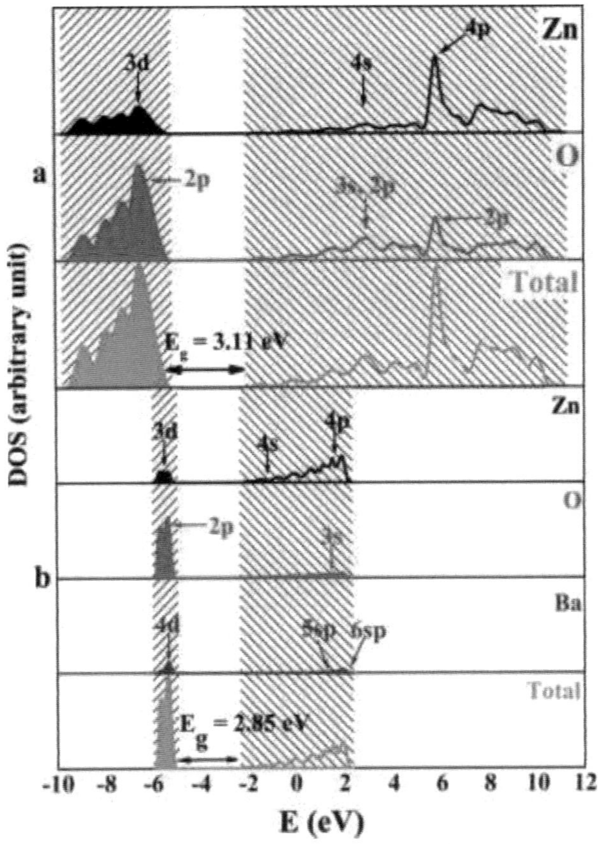

Figure 1. Total and partial projected DOS for (a) ZnO material and (b) ZnO:Ba material at 12.5 %. The hatch range shows the electronic degenerescence of VB and CB from Ba doping.

The energy levels of VB and CB from ZnO are influenced by Ba 4d orbitals, mainly, in band-gap (E_g) region (Figure 1). However, another modification from Ba-doping is an electronic degenerescence of these energy levels on top of VB and bottom of CB from ZnO. It is possible to discuss such degenerescence effect because the projected DOS analysis (Figure 1) was performed from equal number of energy levels as VB as CB, the energy levels number chosen was five. For better visualization of such electronic degenerescence effect was used a hatcher range in Figure 1. Thus, for ZnO, the energy levels as VB as CB occupy an energy range broader than same energy levels on ZnO:Ba (hatcher range in Figure 1). Such electronic effect is important because the electronic structure was changed through low coordination number (CN) of the Ba atom, CN=4 in wurtzite-ZnO; such CN=4 is unusual for Ba, consequently, the orbitals distribution of Ba is also unusual causing a decrease of E_g (Figure 1). For the ZnO (Figure 1a), the theoretical results exhibit a direct E_g of 3.11 eV ($\Gamma - \Gamma$) in according to experimental results [5, 7, 8]. Nevertheless, from doping process (Figure 1b) was calculated two main E_g values and both are lower than E_g calculated for ZnO material. The lowest

E_g for ZnO:Ba was calculated as 2.85 eV refers to an indirect E_g (L – Γ); whereas, a direct E_g (Γ – Γ) was calculated as 3.01 eV. Such E_g values refer to radiation in the visible ultraviolet making of this material a potential alternative to application in electronics devices and photocatalytic process. The main contribution to change the E_g from Ba-doping was caused by degeneration of energy levels belonging to VB and CB of ZnO. More specific, the electronic degenerescence causes an increase of VB energy and a decrease of CB energy.

4. Conclusion

For ZnO and ZnO:Ba materials, the results indicate that Ba-doping at 12.5% in wurtizte-ZnO is a potential alternative to improve the electronic and optical properties of ZnO materials. This improvement is caused by degeneration of energy levels and, consequent reduction of the band-gap.

5. Acknowledgments

The authors acknowledge support from CAPES and Fundação Araucária.

6. References

[1] Kwok H L 1997 *Electronic Materials* (Boston: PWS Publishing Company)

[2] West A R 2006 *Basic solid state chemistry* (Chichester: John Wiley & Sons)

[3] Dhananjay, Nagaraju J and Krupanidhi S B 2006 Effect of Li substitution on dielectric and ferroelectric properties of ZnO thin films grown by pulsed-laser ablation *Journal of Applied Physics* **99**

[4] Xu J, Hao S, Duan X, Gu S and Meng X 2013 Preparation and photoelectric properties of La-doped ZnO films *Journal of Materials Science: Materials in Electronics* **24** 4175-9

[5] Klingshirn C 2007 ZnO: Material, Physics and Applications *ChemPhysChem* **8** 782-803

[6] Gupta M K, Sinha N and Kumar B 2012 Dielectric studies and band gap tuning of ferroelectric Cr-doped ZnO nanorods *Journal of Applied Physics* **112** 014303

[7] Ozgur U, Hofstetter D and Morkoc H 2010 ZnO Devices and Applications: A Review of Current Status and Future Prospects *Proc. IEEE* **98** 1255-68

[8] Özgür Ü, Alivov Y I, Liu C, Teke A, Reshchikov M A, Doğan S, Avrutin V, Cho S-J and Morkoç H 2005 A comprehensive review of ZnO materials and devices *J. Appl. Phys.* **98** 041301

[9] Yang Y C, Song C, Wang X H, Zeng F and Pan F 2008 Giant piezoelectric d33 coefficient in ferroelectric vanadium doped ZnO films *Applied Physics Letters* **92** 012907

[10] Lee S, Lee Y, Young Kim D and Won Kang T 2013 Effects of Ti additives on structural and electric properties of Cr- and Ti-codoped ZnO layers *Journal of Applied Physics* **114** 064102

[11] Heo S, Sharma S K, Lee S, Lee Y, Kim C, Lee B, Lee H and Kim D Y 2014 Effects of Y contents on surface, structural, optical, and electrical properties for Y-doped ZnO thin films *Thin Solid Films* **558** 27-30

[12] Ilican S 2013 Effect of Na doping on the microstructures and optical properties of ZnO nanorods *Journal of Alloys and Compounds* **553** 225-32

[13] Liu L N, Wang C C, Lei C M, Li T, Wang G J, Sun X H, Wang J, Huang S G, Li Y D and Wang H 2013 Relaxor- and phase-transition-like behaviors in ZnO single crystals at high temperatures *Applied Physics Letters* **102**

[14] Hao L, Li Y, Zhu J, Wu Z, Long F, Liu X and Zhang W 2014 Microstructure and memory characteristics of ferroelectric LiNbO3/ZnO composite thin films on Pt/TiO2/SiO2/Si substrates *Journal of Alloys and Compounds* **590** 205-9

[15] Dhananjay, Nagaraju J and Krupanidhi S B 2007 Off-centered polarization and ferroelectric phase transition in Li-doped ZnO thin films grown by pulsed-laser ablation *Journal of Applied Physics* **101** 104104

[16] Chand P, Gaur A and Kumar A 2013 Structural, optical and ferroelectric behavior of hydrothermally grown ZnO nanostructures *Superlattices and Microstructures* **64** 331-42

[17] Glinchuk M D, Kirichenko E V, Stephanovich V A and Zaulychny B Y 2009 Nature of ferroelectricity in nonperovskite semiconductors like ZnO:Li *Journal of Applied Physics* **105** 104101

[18] Jindal K, Tomar M, Katiyar R S and Gupta V 2012 Structural and magnetic properties of N doped ZnO thin films *J. Appl. Phys.* **111** 102805

[19] Gupta M K and Kumar B 2011 High Tc ferroelectricity in V-doped ZnO nanorods *Journal of Materials Chemistry* **21** 14559-62

[20] Singh D P, Gupta S K, Pandey K K, Yadav S P, Varia M C and Manohar R 2013 Ferroelectric liquid crystal matrix dispersed with Cu doped ZnO nanoparticles *Journal of Non-Crystalline Solids* **363** 178-86

[21] Jia C, Chen Y, Guo Y, Liu X, Yang S, Zhang W and Wang Z 2011 Valence band offset of InN/$BaTiO_3$ heterojunction measured by X-ray photoelectron spectroscopy *Nanoscale Research Letters* **6** 1-5

[22] Han T, Meng F Y, Zhang S, Cheng X M and Oh J I 2011 Band gap and electronic properties of wurtzite-structure ZnO co-doped with IIA and IIIA *J. Appl. Phys.* **110**

[23] Kang Y H, Choi J-H, Lee T I, Lee W and Myoung J-M 2011 Thickness dependence of the resistive switching behavior of nonvolatile memory device structures based on undoped ZnO films *Solid State Communications* **151** 1739-42

[24] Lee Y, Lee S, Ryu H T and Kim D Y 2011 Appearance of mulitferroic indications in ZnO:(Cr,Ti) thin films *AIP Conference Proceedings* **1399** 163-4

[25] Srinet G, Kumar R and Sajal V 2014 High Tc ferroelectricity in Ba-doped ZnO nanoparticles *Materials Letters* **126** 274

[26] Haja Hameed A S, Karthikeyan C, Sasikumar S, Senthil Kumar V, Kumaresan S and Ravi G 2013 Impact of alkaline metal ions Mg2+, Ca2+, Sr2+ and Ba2+ on the structural, optical, thermal and antibacterial properties of ZnO nanoparticles prepared by the co-precipitation method *Journal of Materials Chemistry B* **1** 5950-62

[27] Schubert M, Ashkenov N, Hofmann T, Lorenz M, Hochmuth H, v. Wenckstern H, Grundmann M and Wagner G 2004 Electro-optical properties of ZnO-$BaTiO_3$-ZnO heterostructures grown by pulsed laser deposition *Annalen der Physik* **13** 61-2

[28] Voora V M, Hofmann T, Schubert M, Brandt M, Lorenz M, Grundmann M, Ashkenov N and Schubert M 2009 Resistive hysteresis and interface charge coupling in $BaTiO_3$-ZnO heterostructures *Applied Physics Letters* **94** 142904

[29] Voora V M, Hofmann T, Brandt M, Lorenz M, Ashkenov N, Grundmann M and Schubert M 2009 Electrical properties of ZnO–$BaTiO_3$–ZnO heterostructures with asymmetric interface charge distribution *Applied Physics Letters* **95** 082902

[30] Böntgen T, Schöche S, Schmidt-Grund R, Sturm C, Brandt M, Hochmuth H, Lorenz M and Grundmann M 2011 Optical properties of $BaTiO_3$/ZnO heterostructures under the effect of an applied bias *Thin Solid Films* **519** 2933-5

[31] Fan J and Freer R 1997 Varistor properties and microstructure of ZnO–BaO ceramics *Journal of Materials Science* **32** 415-9

[32] Becke A D 1993 Density-functional thermochemistry. III. The role of exact exchange *J. Chem. Phys.* **98** 5648-52

[33] Becke A D 1988 Density-functional exchange-energy approximation with correct asymptotic behavior *Phys. Rev. A* **38** 3098-100

[34] Dovesi R, Orlando R, Civalleri B, Roetti C, Saunders V R and Zicovich-Wilson C-M 2005 CRYSTAL: a computational tool for the ab initio study of the electronic properties of crystals *Zeitschrift für Kristallographie* **220** 571

[35] Dovesi R, Saunders V R, Roetti C, Orlando R, Zicovich-Wilson C M, Pascale F, Civalleri B, Doll K, Harrison N M, Bush I J, D'Arco P and Llunell M 2009 *CRYSTAL09 User's Manual* (Torino: University of Torino)

[36] Peintinger M F, Oliveira D V and Bredow T 2013 Consistent Gaussian basis sets of triple-zeta valence with polarization quality for solid-state calculations *J. Comput. Chem.* **34** 451-9

[37] Hay P J and Wadt W R 1985 Ab initio effective core potentials for molecular calculations. Potentials for K to Au including the outermost core orbitals *J. Chem. Phys.* **82** 299-310

[38] Zagorac D, Doll K, Schön J C and Jansen M 2012 Sterically Active Electron Pairs in Lead Sulfide? An Investigation of the Electronic and Vibrational Properties of PbS in the Transition Region Between the Rock Salt and the α-GeTe-Type Modifications *Chemistry - A European Journal* **18** 10929-36

[39] Pisani C, Dovesi R and Roetti C 1988 *Hartree-Fock ab initio Treatment of Crystalline Systems.*: Springer Berlin Heidelberg)

[40] Kokalj A 1999 XCrySDen—a new program for displaying crystalline structures and electron densities *J. Mol. Graphics Modell.* **17** 176-9

[41] Kokalj A 2003 Computer graphics and graphical user interfaces as tools in simulations of matter at the atomic scale *Comput. Mater. Sci.* **28** 155-68

A dilatometric study of the continuous heating transformations in maraging 300 steel

A G Reis[1], D A P Reis[1,2], A J Abdalla[3], J Otubo[1,2] and H R Z Sandim[4]

[1] Instituto Tecnológico de Aeronáutica, Materiais e Processos, São José dos Campos 12228-900, Brazil
[2] Universidade Federal de São Paulo, Instituto de Ciência e Tecnologia, São José dos Campos 12231-280, Brazil
[3] Instituto de Estudos Avançados, Fotônica, São José dos Campos 12228-001, Brazil
[4] Escola de Engenharia de Lorena – Universidade de São Paulo, Lorena 12602-810, Brazil

e-mail: areis@ita.br

Abstract. Dilatometric study in maraging 300 steel was carried out to study the effect of heating rate on precipitation of intermetallic phases and martensite to austenite transformation. Solution annealed material were subjected to controlled heating-holding-cooling cycles. The martensite to austenite transformation splits into two steps at lower heating rates. The first step enhanced by slow heating rate, occurs through a diffusion process, while the second step, enhanced by a fast heating rate, occurs though a shear process. The extent of precipitation decreases with heating rate, suggesting that precipitation occurs primarily by a diffusional process.

1. Introduction

Maraging steels are strategic materials to the Brazil's technological development due to their high strength combined with high fracture toughness, finding extensive applications in several areas, such as in aerospace and nuclear technologies [1].

Maraging refers to the ageing of martensite. The only transformation that occurs at ordinary cooling rates is martensite formation. The martensite without carbon is quite soft, but heavily distorted by a high dislocation density. Hardening and strengthening of these steels are subsequently produced by heat treating (ageing) for several hours at 480–510°C, caused by precipitation of intermetallic phases such as Ni_3X (X=Ti, Mo) and the formation of a more stable phase Fe_2Mo that demands higher exposure times. During this stage, the metastable martensite in the steels decomposes. Fortunately, the precipitation hardening occurs much more rapidly than the reversion reactions producing austenite and ferrite. Thus, substantial hardening can be produced before reversion occurs. Austenite reversion, or the prevention of it in most cases, is important to ageing, because austenite is a stable phase at room temperature for maraging steel compositions [2-4]. It is noteworthy that both diffusion and shear mechanisms can occur simultaneously during this transformation of austenite reversion. Previous studies have shown that these mechanisms can depend on the heating rates. Dilatometry has been largely used for studies of phase transformations on several types of steels, such as low-carbon steels and maraging steels [5,6].

The present work was carried out on maraging 300 steel using the dilatometric technique to study the effect of heating rate on both the precipitation and the martensite to austenite transformation.

IOP Conf. Series: Materials Science and Engineering **97** (2015) 012006 doi:10.1088/1757-899X/97/1/012006

2. Experimental Procedure

Samples from maraging 300 steel supplied in the solution annealed condition were machined to 4 mm diameter and 18 mm length cylindrical rods. The composition of this steel is listed in table 1.

Table 1. Chemical composition (wt %) of the maraging 300 studied.

Element	Ni	Co	Mo	Ti	Al	C	S	P	Si	Mn	Fe
wt %	19.00	9.37	4.94	0.63	0.08	0.008	0.002	0.004	0.06	0.01	Balance

Heating and cooling cycles were performed in the quenching Linseis-L75V 1400 RT dilatometer under argon atmosphere. For these tests, heating was carried out with heating rates of 10 and 20°C/min. Length increase (ΔL) and temperature (T) changes were measured during heating for each heating rate.

3. Results and Discussion

A typical dilatometric curve of the maraging 300 steel tested at a heating rate of 10 °C/min is shown in figure 1.

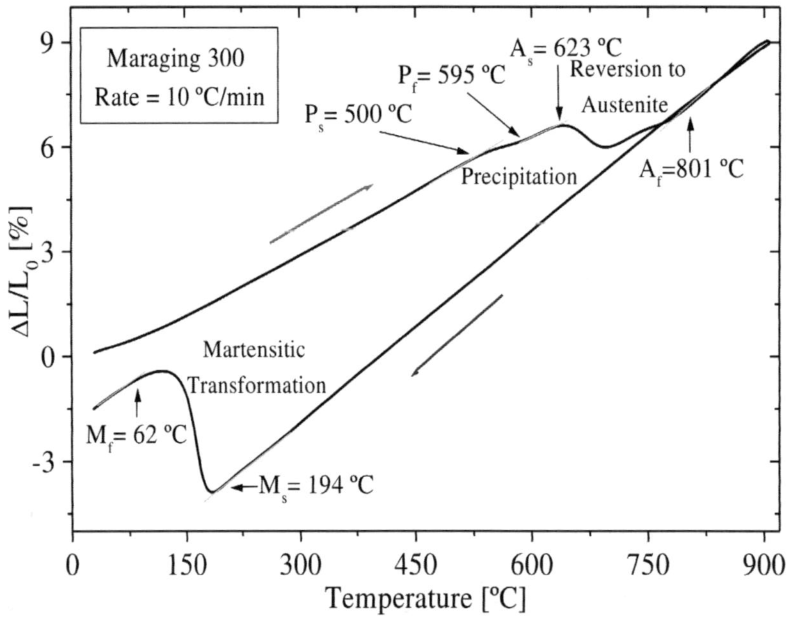

Figure 1. Dilatometric curve of heating and cooling for Maraging 300, heating rate of 10 °C/min.

Uniform expansion continues until 500 °C when a small contraction starts to occur, indicating the start of precipitation at this temperature (P_s). This is followed by a small amount of linear expansion with increasing temperature to 595 °C (P_f). When the temperature is raised to 623°C, a large contraction appears in the dilatometry curve, which can be taken as the start of the austenite formation (A_s). At 720 °C, the contraction starts to slow down, until 801°C when the curve resumes linear expansion. Therefore, the austenite transformation ends at around 801°C (A_f). Complete solution is ensured by heating continuously to 900°C and holding at this temperature for 1 minute. During cooling to room temperature, there is drastic expansion at approximately 194 °C, owing to the sudden start and the rapid transformation from austenite to martensite, corresponding to the martensite start temperature (M_s). The martensite finish temperature (M_f) is about 62 °C. Thus, this maraging 300 steel should have a single-phase structure upon cooling to room temperature that is martensite.

Figure 2 presents the variation of the slope of dilatometric curves as a function of temperature to the heating rates of 10 and 20 °C/min for the maraging 300 studied.

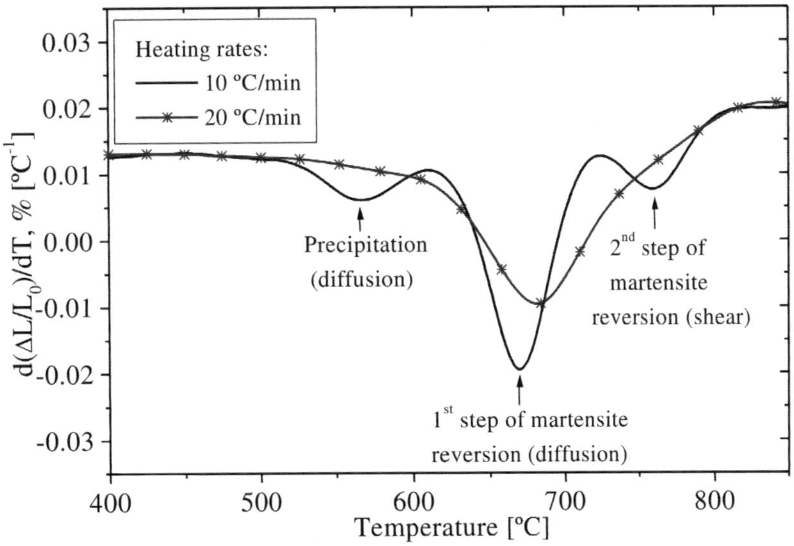

Figure 2. Slope of dilatometric curves for maraging 300 at heating rates of 10 °C/min and 20 °C/min indicating the precipitation and martensite to austenite transformation.

It is observed from figure 2 that at the lower heating rate, the austenite transformation splits in two steps. At a lower heating rates, the martensite starts to transform to austenite by a diffusion process and at sufficiently high temperatures (>720 °C) the thermodynamic driving forces takes over, thus transforming the remaining martensite to austenite by shear mechanism. Since at high heating rates there is insufficient time for diffusion to take place, therefore the entire martensite transforms to austenite by a single step shear mechanism [5,6].

It´s also possible to observe from figure 2 that as the heating rate increases, the extent of precipitation decreases. This result indicates that precipitation in maraging 300 is primarily diffusion-controlled. For higher heating rates the time available for precipitation decreases leading to a smaller specimen contraction, due to the smaller volume fraction of precipitates formed during heating at higher heating rates [5,6].

4. Conclusions

The martensite to austenite transformation in maraging 300 steel shows a tendency to split into two steps as the heating rate is decreased. The first step of this transformation, enhanced by a slow heating rate, occurs through a diffusion process, while the second step, enhanced by a fast heating rate, occurs through a shear process.The extent of precipitation decreases with increasing heating rate, suggesting that precipitation occurs primarily by a diffusional process.

Acknowledgments
The authors are grateful to CNPq (Proc.nº 141274/2013-1), CAPES (Proj. Pró-Defesa 014/08) and FAPESP for the financial support.

References

[1] Assunção F C R 2010 *Materiais Avançados 2010-2022* (Brasília: Centro de Gestão e Estudos Estratégicos).
[2] Sha W and Guo Z 2009 *Maraging steels - Modelling of microstructure, properties and*

applications (Cambridge, Woodhead Publishing Limited)
[3] Hall A M and Slunder C J 1968 *The metallurgy, behavior, and application of the 18-percent Nickel Maraging steels – A survey* (Battelle Memorial Institute, NASA SP-5051)
[4] Rohrback K and Schmidt M 1991 ASM Handbook. Properties and Selection of Irons Steels and High Performance Alloys. In: Maraging Steels vol 1 ed ASM HANDBOOK COMMITTEE pp 1870-1887
[5] Kapoor R, Kumar L and Batra I S 2003 *Mat. Sci. Eng. A.* **352** 318-24.
[6] Carvalho L G, Andrade M S, Plaut R L, Souza F M and Padilha A F 2013 *Mater. Res.* **16-4** 740-4.

IOP Conf. Series: Materials Science and Engineering **97** (2015) 012007 doi:10.1088/1757-899X/97/1/012007

Photoelectrochemical study of Ta_3N_5 nanotubes for water splitting

Sherdil Khan[1], Marcos J. Leite Santos *[2], Jairton Dupont [2], Sergio R. Teixeira*[1]

[1] Instituto de Física, Universidade Federal do Rio Grande do Sul, Av Bento Gonçalves 9500 POBox - 15051 CEP 91501-970, POA-RS, Brazil.
[2] Instituto de Química, Universidade Federal do Rio Grande do Sul, Av Bento Gonçalves 9500 POBox 15003, CEP 91501-970, POA-RS, Brazil.

E-mail address: *srgrbrtxr@gmail.com*

Abstract. Nanotubes (NTs) of Ta_3N_5 were synthesized by nitridation of Ta_2O_5 NTs. The samples were studied by scanning electron microscopy, UV-VIS spectrophotometry, x-ray diffraction and photoelectrochemical (PEC) measurements carried out in aqueous solutions of Na_2SO_4 and $Fe(CN)_6{}^{3-/4-}$. The results show the presence of trapping states on illuminated pristine Ta_3N_5 NTs in Na_2SO_4 (aq). These trapping states act as recombination center for photogenerated holes, affecting the photocatalytic performance of Ta_3N_5 NTs. On the other hand, by using $Fe(CN)_6{}^{3-/4-}$ (aq) the photogenerated holes were scavenged efficiently without giving rise to the trapping states at Ta_3N_5 NTs/solution interface. The results obtained by cyclic voltammetry, linear sweep voltammetry and electrochemical impedance spectroscopy have shown that the presence of these trapping states is a limiting step for water oxidation using pristine Ta_3N_5 NTs.

1. Introduction

Aiming to produce hydrogen from efficient solar-driven water splitting systems, many research groups have devoted their efforts to the study and development of photocatalysts [1,2, 3]. Several metal oxide semiconductors have been applied in photocatalytic and photoelectrochemical (PEC) water splitting, however most of these materials present large band gaps (> 3.0) absorbing light only within the ultraviolet region [4]. Once the visible spectrum accounts nearly half of the total energy from the sun that reaches the surface of the earth, the development of stable materials presenting absorption within the visible range is a key factor towards efficient water splitting systems. Within this context, Ta_3N_5 has emerged as a promising candidate, presenting adequate bandgap of ca. 2.1 eV and band edge at ca. 600 nm [5, 6]. According to the literature, the PEC water splitting performance of Ta_3N_5 is highly dependent on the loading of a suitable co-catalyst, while pristine Ta_3N_5 presents poor PEC activity [7]. Nevertheless, improved PEC activity from pristine Ta_3N_5 can be obtained by using a hole scavenging solution of $Fe(CN)_6{}^{3-/4-}$ [8]. In fact, the use of hole scavenging electrolytes such as Na_2SO_3, $Fe(CN)_6{}^{3-/4-}$, have resulted to improve PEC performance of other semiconductors [2,9].

The mechanism behind the poor PEC performance of pristine Ta_3N_5 based PEC devices, in water oxidation conditions, is not fully understood and a throughout study of the Ta_3N_5/electrolyte interface is necessary. Hence, in this work we have studied the PEC water splitting of Ta_3N_5 nanotubes. Firstly; we synthesized Ta_2O_5 NTs by modified anodization which helped us to preserve the tubular morphology at high temperatures. Later, the precursor Ta_2O_5 NTs were transformed to Ta_3N_5 NTs by thermal treatment under ammonia environment. We have compared the PEC activity of Ta_3N_5 photoanodes under water oxidation conditions using Na_2SO_4 (aq) and hole scavenger oxidation using $Fe(CN)_6{}^{3-/4-}$(aq).

2. Experimental procedure

Ta (99.95%) sheet was cut into 2 cm × 2 cm small pieces, cleaned with acetone in ultrasound bath for 30 min, rinsed with water and dried under nitrogen flux. The anodization was performed in H_2SO_4 + 1 vol% of HF and 4 vol% of H_2O at 10°C during periods of 5 min and 20 min in a two electrode electrochemical cell. Ta disks were used as anode and Cu disks as cathode. The applied voltage was fixed to 50 V ramped by 10 V/s. Nitridation was performed at 800°C and 900°C (heating and cooling rate: 5°C/min) for 3 h under 100 ml/min of NH_3 flux. Photoelectrochemical measurements were performed separately in 0.1 M Na_2SO_4 (pH, 7.85), 0.1 M $K_4[Fe(CN)_6$ and 0.1 mM $K_3[Fe(CN)_6]$ (pH, 7.5). Ta_3N_5 NTs were used as working electrode, Pt wire as counter electrode and Ag/AgCl was used as a reference. The Ta_3N_5 NTs samples were characterized by scanning electron microscopy (SEM) performed in an EVO 50 from Zeiss, operated at 10 kV. Absorption spectra in the UV-Vis were obtained using a CARY 5000, Grazing angle X-ray diffraction (XRD) patterns were recorded by Shimadzu, Maxima XRD-7000 diffractometer with an incident angle of 0.3° with Cu K_α radiation (λ = 1.54 Å) at a 2θ range from 15° to 70° with a 0.05° step size and measuring time of 5s per step. Cyclic voltammetry (CV), Linear Sweep Voltammetry (LSV) and Electrochemical Impedance Spectroscopy (EIS) were performed by Auto-lab (AUT 84503) potentiostat. A Xe lamp (300W) was used as a source of electromagnetic radiation and an AM 1.5 filter was used, to obtain simulated solar spectrum. The light intensity was calibrated for 1 Sun conditions (100 mW.cm^{-2}), using a silicon photodiode.

3. Results and discussion

Fig. (1a, b) show the SEM images of the Ta_3N_5 NTs, displaying open top and closed bottom tubes. The as-anodized Ta_2O_5 NTs (not shown) were ca. 1.6 µm long, presenting wall thickness of 34 nm and outer diameter of 130 nm. In a previous study on Ta_2O_5 NTs prepared by anodization, we have found that anodization temperature controls the adherence of the NTs on the substrate. In addition to improve adherence, lowering the anodization temperature results in thicker walled nanotube [4].

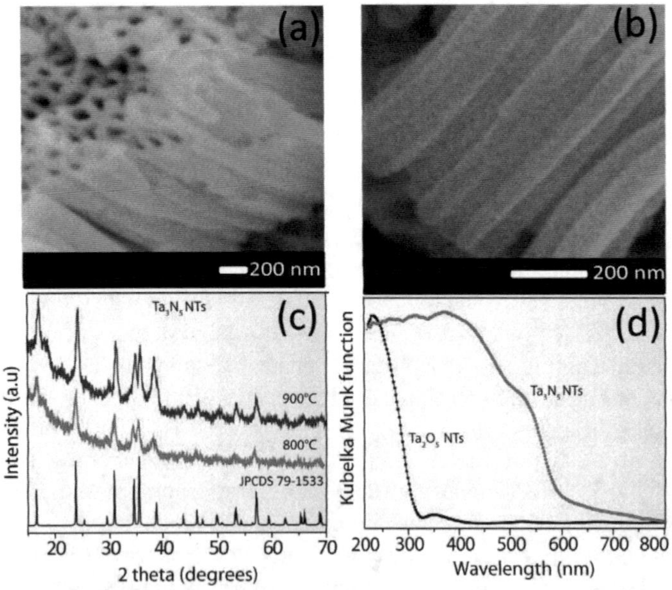

Fig. 1 SEM images of Ta_3N_5 NTs synthesized at (a) 800°C and (b) 900°C for 3 h. (c) XRD patterns of Ta_3N_5 NTs and (d) UV-Vis spectra of Ta_2O_5 and Ta_3N_5.

In this study, thick walled and well adhered Ta_2O_5 NTs were obtained at 10°C, making them preserve their morphology even when heated at 900°C. After nitridation at 800°C and 900°C, the length of the Ta_3N_5 NTs were reduced to ca 1.4 μm and 1.1 μm, respectively; compared to the length of the as-anodized NTs. In addition, slight decrease in the diameter of the NTs was observed. These changes in dimension are related to the density difference between the Ta_2O_5 and Ta_3N_5 structures [10]. Fig. 1c presents the grazing angle XRD patterns of the samples obtained by angle of incidence of 0.3°, with respect to the substrate surface. All peaks match with the orthorhombic phase of Ta_3N_5 (JPCDS file 79-1533). By using Rietveld Refinement technique, the average grain size of the nanotubes synthesized at 800°C and 900°C were 19.71, 43.51 nm respectively. Fig. 1d shows the absorption spectra of the as-anodized Ta_2O_5 and the Ta_3N_5 obtained at 800°C. After nitridation one can observe a large red-shift of the band edge from 320 nm (Ta_2O_5) to 600 nm (Ta_3N_5) and bandgap values of 3.9 and 2.1 eV respectively. In addition, the spectrum of Ta_3N_5 display a broad absorption mode above ca. 600 nm, which is related to ionic defects in the lattice resulting from the nitridation process [11].

In order to compare our findings to the literature, we have focused our study on the Ta_3N_5 samples obtained at 800°C [12]. Fig. (2a, b) display the cyclic voltammograms of the sample nitrided at 800°C for 3 h. The experiments were performed under dark (Fig. 2a) and under illumination (1 Sun) (Fig. 2b) in Na_2SO_4 (aq) electrolyte. For CV obtained under dark (Fig. 2a), as the potential is swept toward negative values, a strong increase of the capacitive current can be observed, starting from ca. −0.2V. Under cathodic applied potential the chemical capacitance associated with the electron population on the semiconductor begins to dominate the Helmholtz capacitance, where the quasi-Femi level gets closer to the conduction band which leads to a capacitive current [13]. The un-coordinated metallic ionic species that exist on the surface, will give rise to electronic states below the conduction band where capacitance is controlled by the Helmholtz region.

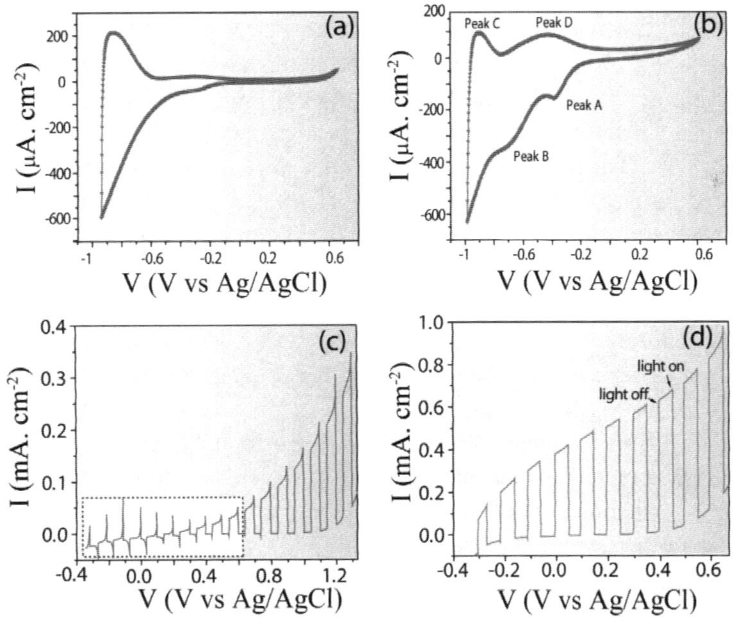

Fig. 2 CV curves of Ta_3N_5 NTs under (a) dark and (b) illumination. LSV curves of prisitine Ta_3N_5 NTs in aqueous electrolytes of (c) Na_2SO_4 and (d) $Fe(CN)_6^{3-/4-}$.

Considering the cathodic scan, under potentials more negative than −0.2 V vs Ag/AgCl, an exponential rise of the current is observed, which turned into an anodic peak at ca. −0.8 V, after the scan direction of the voltammetry is reversed from cathodic to anodic. The exponential increase in cathodic current gives rise to an exponential capacity, which is related to a charge accumulation on the semiconductor [13]. Furthermore, from ca. −0.2 towards positive potentials a constant current can be observed. Nevertheless, whether the scan is anodic or cathodic the current in that potential window is constant. This behavior can be related to the carrier depletion on Ta_3N_5 NTs which is characteristic of the Mott-Shottky window. Fig. 2b displays the cyclic voltammetry curves of Ta_3N_5 NTs obtained under illumination (1 Sun). The exponential rise of current under negative potentials shown in Fig. 2a is transformed into peaks. For the pH used in the current study, the water reduction potential is −0.66V vs Ag/AgCl; therefore, the peak B which appears at ca. 0.7V vs Ag/AgCl and C are related to the redox process of water. An interesting feature of the CV under illumination is peak A appearing at ca. −0.3V vs Ag/AgCl, which cannot be related to any redox process of the electrolyte. Furthermore, upon returning the cathodic scan to the anodic direction a wider peak (peak D) appears. Similar features have been observed in the literature for hematite, where a sharp cathodic peak was related to the filling of narrow deep trapping states and broader anodic peak corresponded to relatively slow depopulation of these states [14].

In order to gain further insight on the role of trapping states, we have obtained LSV curves of pristine Ta_3N_5 NTs synthesized at 800°C during 3h. The experiments were performed at a scan rate of 10 mV/s, by chopping the light under a frequency of 0.2 Hz, using Na_2SO_4 (aq) Fe $(CN)_6^{3-/4-}$ (aq) redox couple Fig (2c, d). The PEC water splitting potential window where the solar to hydrogen efficiency can be obtained for Na_2SO_4 (aq) system is highlighted by black color dotted line in Fig. 2c. The photocurrent obtained in Fe $(CN)_6{}^{3-/4-}$ (aq) is clearly higher than in Na_2SO_4 (aq) for that potential window. Even when higher biasing is applied for Na_2SO_4 (aq) the photocurrent could not be improved as compared to Fe $(CN)_6{}^{3-/4-}$ (aq). In addition, the LSV curve of the system using $Fe(CN)_6{}^{3-/4-}$ (aq) redox couple presented better fill factor. In Na_2SO_4 (aq), at lower potentials, one can observe “overshoot” current spikes decaying momentarily and upon turning the light off sharp cathodic spikes are also observed. These results strongly suggest that under illumination, photogenerated holes reaching the Ta_3N_5 NTs surface in contact with Na_2SO_4 (aq), will be trapped and there will be large recombination of the photogenerated carriers. Once light is turn off, the holes which are trapped are oxidized resulting in a sharp cathodic peak. On the other hand, for higher applied potential biasing the cathodic spikes cannot be observed in Na_2SO_4. In these applied potentials the strong electric field at the double layer at the interface will force to neutralize the holes, as well as, it will separate the photogenerated charge carriers. These results clearly indicate that in order to obtain improved PEC performance from Ta_3N_5 NTs, the holes trapped on the surface are only neutralized by applying higher biasing potential. That is a major drawback, once the use of a large potential biasing will reduce the solar energy efficiency conversion for water splitting. Interestingly, the improved photocurrent and the absence of the cathodic spikes from Ta_3N_5 NTs in Fe $(CN)_6{}^{3-/4-}$ strongly suggests that the efficient hole scavenging nature of Fe $(CN)_6{}^{3-/4-}$ do not give rise to the trapping states.

To elucidate the mechanism of charge transfer across the Ta_3N_5 NTs/Na_2SO_4-interface, Nyquist plots were obtained at different applied potentials (Fig. 3a). Fig. 3a shows incomplete semicircles characteristic of a non-ideal capacitance behavior [15]. The data were fitted by an equivalent circuit shown in Fig. 3b [8]. The observed decrease in charge transfer resistance upon increasing the biasing potential, corroborates the earlier suggestion that; it is necessary to apply higher biasing to help to neutralize the trapping states. EIS corroborates these findings showing that for low charge transport resistance, higher biasing should be applied externally. In addition, from the observed decrease in charge transfer resistance upon increasing the biasing potential; we may infer that the fill factor of Ta_3N_5 NTs in Na_2SO_4 (aq) would be poor which can clearly be seen in Fig. 2c.

For the analyses discussed above we have used Ta_3N_5 NTs obtained for 20 min of anodization and nitrided at 800°C for 3 hours. In order to investigate possible effect of the length of the NTs on the trapping states and to improve the photocurrent, we have decreased the length by anodizing the samples for only 5 min and further nitriding them at 800°C during 3 hours. The LSV curve obtained for this sample is shown in Fig. 3c (red color solid line).

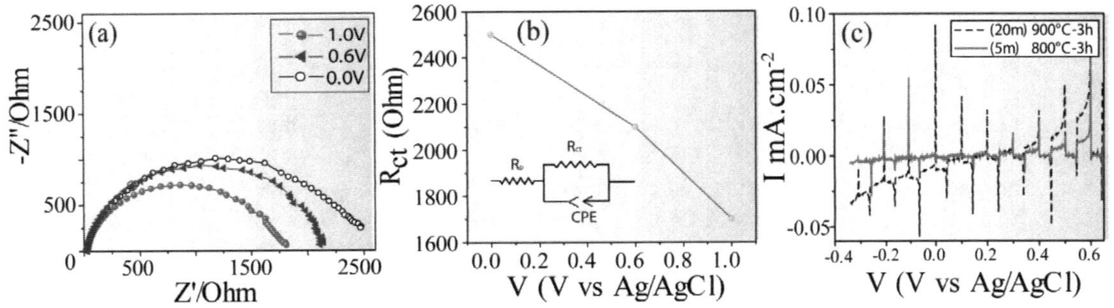

Fig.3 Nyquist plots of Ta_3N_5 NTs (a) and equivalent circuit fitting; charge transfer resistance R_{ct} vs applied potential (b). LSV curves of Ta_3N_5 NTs synthesized under two different conditions (c).

Comparing the photocurrent obtained from the sample with the longer length NTs (Fig. 2c, highlighted potential window), no improvement can be observed. Therefore, short length NTs, which are expected to present lower photogenerated charge carrier resistance, could not compete with the trapping states. To evaluate if the nanocrystallinity of the NTs may cope with the trapping states, we have obtained LSV curves of the sample anodized for 20 min and then nitrided at 900°C for the same period of 3 hours (Fig. 3c - black color dotted line). The cathodic and anodic spikes can be observed clearly. In addition, photocurrent is still not improved. Hence the trapping states are independent of the length and the crystallinity of pristineTa_3N_5 NTs. In the literature, the best PEC activity for Ta_3N_5 was observed by Domen´s group, once Ta_3N_5 was doped with Ba firstly; and then loaded with Co-Pi as a co-catalyst [7]. We may infer that the improvement observed in that report is related to avoiding the trapping states. Hence, we strongly suggest, to improve PEC performance of Ta_3N_5 NTs a similar kind of methodology should be followed.

4. Conclusions

In this work we have observed that photoelectrochemical activity of Ta_3N_5 NTs is highly dependent on the hole scavenging from the surface of the nanotubes. To improve water oxidation in Na_2SO_4 (aq) a higher applied potential is required, which is a major drawback to solar energy efficiency conversion of water splitting. On the other hand, due to the efficient hole scavenging ability of $Fe(CN)_6{}^{3-/4-}$ (aq), low biasing potential is required for Ta_3N_5 NTs that has presented an improved PEC performance once in this electrolyte no trapping state was observed.

Acknowledgements

The authors are grateful for financial support from the following Brazilian agencies: Conselho Nacional de Desenvolvimento Científico e Tecnológico (Processo: 477804/2011-0,

490221/2012-2 and 472243/2013-6) and Fundação de Amparo à Pesquisa do Estado do Rio Grande do Sul (FAPERGS – PqG 2011 Proc. 11/0837-0).

References

[1] Fernandes J A, Migowski P, Fabrim Z, Feil A F, Rosa G, Khan S, Machado G J, Fichtner P F P, Teixeira S R, Santos M J L and Dupont J 2014 TiO2 nanotubes sensitized with CdSe via RF magnetron sputtering for photoelectrochemical applications under visible light irradiation *Phys. Chem. Chem. Phys.*

[2] Klahr B, Gimenez S, Fabregat-Santiago F, Bisquert J and Hamann T W 2012 Electrochemical and photoelectrochemical investigation of water oxidation with hematite electrodes *Energ Environ Sci* **5** 7626.

[3] Goncalves R V, Migowski P, Wender H, Feil A F, Zapata M J M, Khan S, Bernardi F, Azevedo G M and Teixeira S R 2014 On the crystallization of Ta2O5 nanotubes: structural and local atomic properties investigated by EXAFS and XRD *Crystengcomm* **16** 797-804.

[4] Goncalves R V, Migowski P, Wender H, Eberhardt D, Weibel D E, Sonaglio F C, Zapata M J M, Dupont J, Feil A F and Teixeira S R 2012 Ta2O5 Nanotubes Obtained by Anodization: Effect of Thermal Treatment on the Photocatalytic Activity for Hydrogen Production *J. Phys. Chem. C* **116** 14022-30.

[5] Chun W-J, Ishikawa A, Fujisawa H, Takata T, Kondo J N, Hara M, Kawai M, Matsumoto Y and Domen K 2003 Conduction and Valence Band Positions of Ta2O5, TaON, and Ta3N5 by UPS and Electrochemical Methods *The Journal of Physical Chemistry B* **107** 1798-803.

[6] Moriya Y, Takata T and Domen K 2013 Recent Progress in the Development of (Oxy)nitride Photocatalysts for Water Splitting under Visible-Light Irradiation *Coordination Chemistry Reviews*.

[7] Li Y, Zhang L, Torres-Pardo A, Gonzalez-Calbet J M, Ma Y, Oleynikov P, Terasaki O, Asahina S, Shima M, Cha D, Zhao L, Takanabe K, Kubota J and Domen K 2013 Cobalt phosphate-modified barium-doped tantalum nitride nanorod photoanode with 1.5% solar energy conversion efficiency *Nature Communications* **4.**

[8] Kado Y, Lee C-Y, Lee K, Müller J, Moll M, Spiecker E and Schmuki P 2012 Enhanced water splitting activity of M-doped Ta3N5 (M = Na, K, Rb, Cs) *Chem Commun* **48** 8685.

[9] Zhang Q, Celorrio V, Bradley K, Eisner F, Cherns D, Yan W and Fermín D J 2014 Density of Deep Trap States in Oriented TiO2 Nanotube Arrays *The Journal of Physical Chemistry C* **118** 18207-13.

[10] Zhang Q H and Gao L 2004 Ta3N5 nanoparticles with enhanced photocatalytic efficiency under visible light irradiation *Langmuir* **20** 9821-7.

[11] Dabirian A and van de Krol R 2013 Resonant optical absorption and defect control in Ta3N5 photoanodes *Appl Phys Lett* **102.**

[12] Cong Y, Park H S, Wang S, Dang H X, Fan F-R F, Mullins C B and Bard A J 2012 Synthesis of Ta3N5 Nanotube Arrays Modified with Electrocatalysts for Photoelectrochemical Water Oxidation *J. Phys. Chem. C* **116** 14541-50.

[13] Bisquert J, Fabregat-Santiago F, Mora-Seró I, Garcia-Belmonte G, Barea E M and Palomares E 2008 A review of recent results on electrochemical determination of the density of electronic states of nanostructured metal-oxide semiconductors and organic hole conductors *Inorg. Chim. Acta* **361** 684-98.

[14] Bertoluzzi L, Badia-Bou L, Fabregat-Santiago F, Gimenez S and Bisquert J 2013 Interpretation of Cyclic Voltammetry Measurements of Thin Semiconductor Films for Solar Fuel Applications *J Phys Chem Lett* **4** 1334-9.

[15] Pajkossy T 1994 Impedance of rough capacitive electrodes *Journal of Electroanalytical Chemistry* **364** 111-25.

IOP Conf. Series: Materials Science and Engineering **97** (2015) 012008 doi:10.1088/1757-899X/97/1/012008

Sensor development exploiting graphite-epoxy composite as electrode material

André L M Azevedo[1,3], Renato S Oliveira[2], Eduardo A Ponzio[1], Felipe S Semaan[1]

[1]Instituto de Química, Universidade Federal Fluminense, Outeiro São João Batista, sn, Centro, Niterói, Rio de Janeiro, Brazil, 24020-141
[2]Instituto Federal de Educação, Ciência e Tecnologia do Rio de Janeiro, Rua Dr. José Augusto Pereira dos Santos, Neves, São Gonçalo, Brazil, 24425-004.

[3] To whom correspondence should be addressed: andreazevedo@id.uff.br

Abstract. This study presents some results regarding the development and characterization of graphite-epoxy composites for use as working electrodes in electroanalysis. Such composites were preliminary assessed by TGA-DTA, AFM, XDR and cyclic voltammetry (CV), standing for a suitable stable and low cost material for electroanalytical purposes. The described material was used, in its best proportion (65% graphite m/m), to build a cell electrochemistry.

1. Introduction

Since the incipient and historical application of Sir Humphrey Davy[1], carbon electrodes have been extensively exploited for both electrochemical and electroanalytical purposes. Despite this, however, with the advent of polarography, many reducible analytes, including metal ions, could be better determined by using mercury as a working electrode. Observing the wide range of oxidizable compounds and considering the limitations of mercury for such analysis Adams, in 1958, revisited some of first carbon-based electrodes and proposed the use of carbon-insulator dispersions as electrode materials, a new material capable of complementing the existing technique[2].

Composites are hybrid materials consisting of a conducting phase, usually graphite powder, mixed in different proportions with an insulating phase, being such product homogeneous on macroscopic level but with clearly defined zones in microscopic dimensions[3]. Mechanical characteristics of the material are often controlled by the insulating phase selected, since the conductive phase generally employed is graphite powder. Among the insulating materials used, some should be highlighted: organic solvents[2], mineral oil[4], paraffin[5], poly(4-vinyl pyridine)[6], epoxy [7] and polyurethane resins[8], [9]. Generally solid composite electrodes exhibits higher mechanical stability than those found for paste-like electrodes[10], fundamental aspect when building robust and reproductive sensing systems. From the simplest to the most sophisticated applications, there are many good examples of how useful are graphite-epoxy composites: starting from the determination of metals[11], covering many biochemical and pharmaceutical analysis such as the simultaneous determination of dopamine and ascorbic acid[12], melamine[13], atenolol[14], and urea[7], reaching the recent developments in electronic tongue systems[15].Polymers used can be incorporated by simply dissolving in a corresponding solvent[10] or the polymerization can be carried out *in situ*[9].

2. Experimental procedures

One disposable electrochemical cell comprising working electrode, counter electrode of stainless steel, silver conductive epoxy reference (pseudo-reference) electrode was constructed. The electrodes were mounted on a plunger of a 5 mL syringe, being then the syringe filled by insulating polymers for mechanical purposes (Figures 1 and 2).

The working electrode was a graphite-epoxy composite prepared through the conventional method described by Swofford e Carman[16], being the polymerization carried out *in situ*, by mechanical dispersion of suitable amounts of graphite powder (Sigma-Aldrich, USA, 2-20 μm) in epoxy resin (Avipol, Brasil, Silaex SQ 2126-3024); the system was left to cure for 24 hrs under pressure and polished with suitable tools. A copper wire was inserted through the composite to establish electrical contact. The frame utilized to construct was based on an insulin syringe 0.50 mL with 3.1 mm internal diameter. Different compositions were prepared (55-80%, w/w graphite content) and characterized by thermogravimetric analysis (TGA-DTA), atomic force microscopy (AFM), X-ray diffraction (XDR) and cyclic voltammetry (CV) according to current-potential responses (5.0×10^{-3} mol L^{-1} $K_3[Fe(CN)_6]$ in 0.5 mol L^{-1} KCl)[17], showing suitable long-term stability.

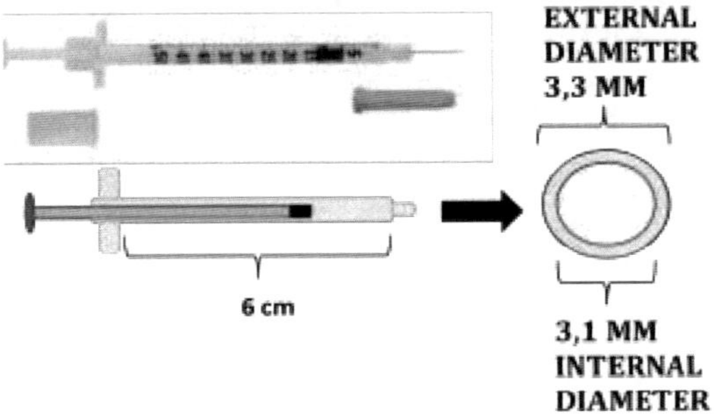

Figure 1. Body of working electrode

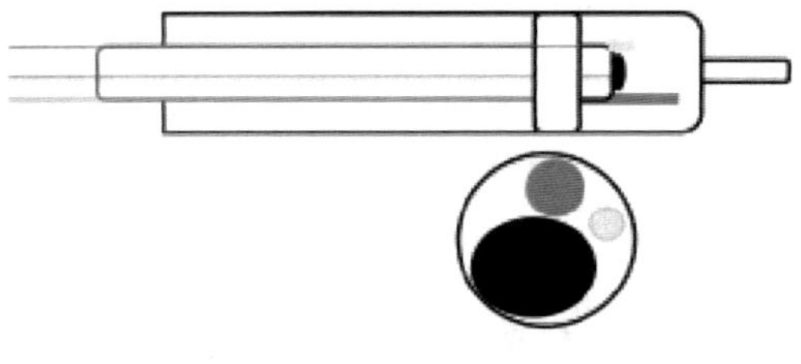

Figure 2. Arrangement of electrodes in electrochemical cell

3. Results and discussion

3.1. Preliminary characterization

The results revealed that the material has a thermal stability up to 150°C, besides this, no chemical relevant interaction between graphite and epoxy was observed (Figure 3). By means AFM studies and calculation, main qualitative and quantitative aspects such as topography and roughness (Figure 4) were assessed and taken into account during global evaluation of electrode performance and stability.

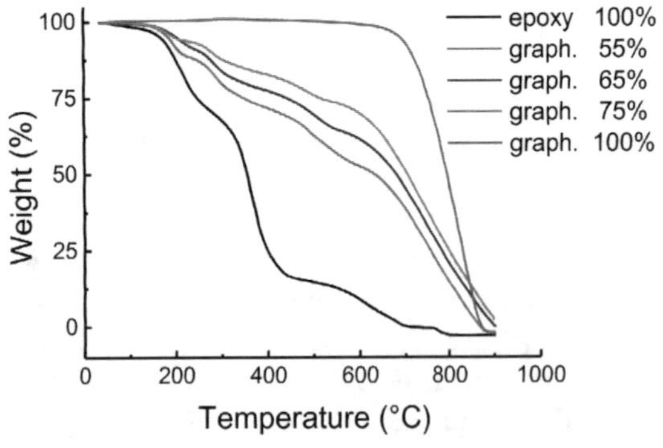

Figure 3. Thermal analysis of different compositions

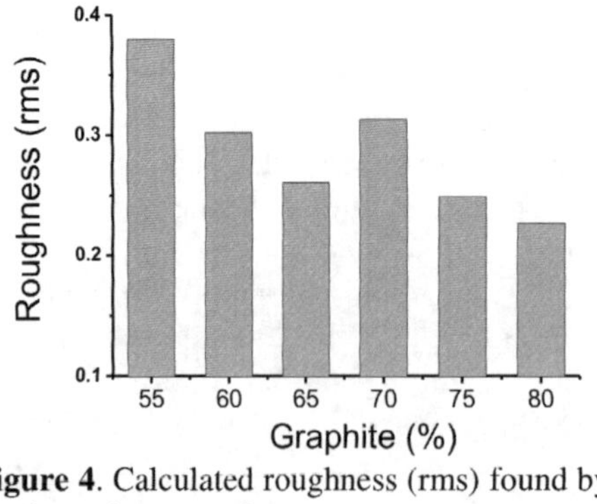

Figure 4. Calculated roughness (rms) found by AFM results and topography

XRD measurements reveal that even with adding an amorphous compound, graphite phase with hexagonal crystal system is maintained (Figure 5)[18]. Although fast and simple, the procedures used for surface polishment and cleaning do not compromise the surface homogeneity, as presented by Figure 6; experiments carried out by AFM provided a roughness (RMS) of circa 0.261. Electroactive areas for composite electrodes (55 to 80%) were determined by cyclic voltammetry. Such studies aimed to evaluate the composition of best electrochemical performance, through a well-known redox probe (Figure 7).

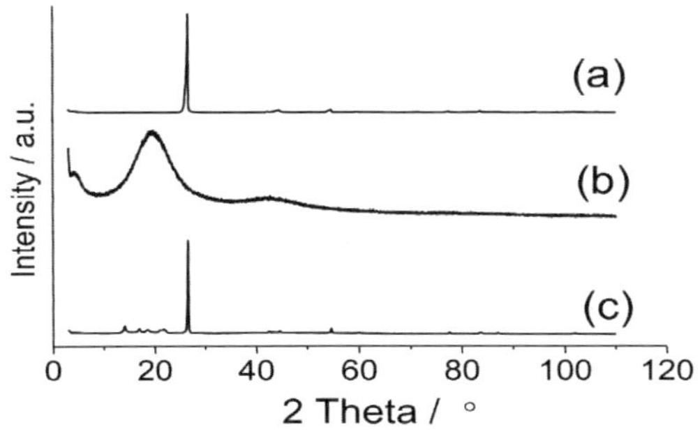

Figure 5. XRD for graphite powder in (a), epoxy in (b) and grahite-epoxy 65% composite in (c)

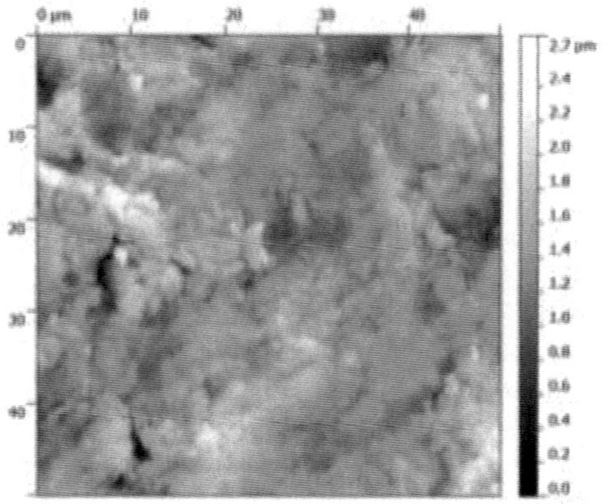

Figure 6. Surface of graphite-epoxy 65% composite by AFM.

3.2. *Electroanalytical performance*

The composite electrodes showed reliable responses, with linear dependence between both currents, cathodic and anodic, and scan rates assessed, as expected. Best results for stability, mechanical and electrical properties were found to be 65% (w/w) (64.4% graphite, characterized via TGA), being these close to the previously described by Trijuque et al.[19] (62% graphite). Potential window and applicability in diverse pH media were also assessed and compared to a commercial glassy carbon electrode with similar geometric surface, leading to equivalent results, but with much lower costs. Besides these, the obtained surface showed to be excellent for chemical modifications aiming to increasing sensitivity, and/or selectivity, which are being currently developed by our group in many different fields.

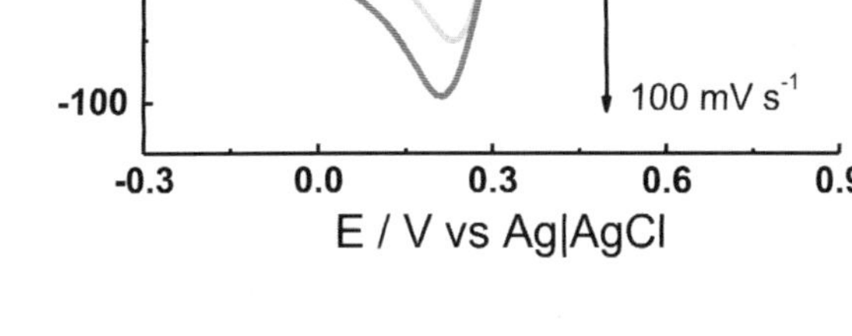

Figure 7. Electrochemical performance assessed by cyclic voltammetry in different scan rates, using outer sphere probe (5.0×10^{-3} mol L^{-1} $K_3[Fe(CN)_6]$ in 0.5 mol L^{-1} KCl)

4. Concluding remarks

A suitable stable and low cost material for electroanalytical application is hereby described. Its hardness facilitates polishing, and subsequently recovery of the surface, allowing stable but transient chemical modifications as well as renewal. The methodology of polishing on abrasive surface resulted on low roughness topography. Economically such material showed to be easy to build and cost-effective when compared to glassy carbon commercially available sensors. For analytical purpose such composite is being evaluated as substrate for gold-nanoparticles immobilization in chitosan-modified and cellulose acetate films, with good preliminary results. The electrochemical cell developed can be directly used to determinate many inorganic and/or organic compounds in real samples or even undergoes to chemical modification to more specialized situations.

References

1. Davy H and Davy J 1838 *The Collected Works Or Sir Humphry Davy: Memoirs of the life of Sir Humphry Davy, by his brother, John Davy*. Smith, Elder and Company.
2. Adams R N 1958 *Anal. Chem.*, **30** (9), 1576
3. Milton G W 2002 *The Theory of Composites*. Cambridge University Press.
4. Kalcher K. 1990 *Am Chem Soc (ACS)*. **419,** 33.
5. Matson W R 1974 *Electrode for anodic stripping voltammetry*. US3855099A.
6. Geno P W et al. 1985 *Jour of Elect Chem and Inter Elect.* **183** (1–2), 155
7. Alegret S et al 1993 *Sens and Act B: Chem* **16** (1–3), 453
8. Toledo D et al 2005 *Anal. Bioanal. Chem.* **381,** 1161
9. Anderson J E, Tallman D E 1976 *Anal. Chem.*, **48,** 209
10. Diewald W et al 1994 *Analyst* (Cambridge, U. K.) **119,** 299
11. Ly S Y et al 2007 *Anal. Lett.* **40,** 2683
12. Ilinoiu E et al 2013 *Sensors* **13** (6), 7296.
13. Zhu Y et al 2011 *Am. Jour. of Anal. Chem.* **2** (5), 6
14. Cavalheiro et al 2011 *Int. Jour. of Env. Anal. Chem.* **92** (5), 561
15. Cetó X et al 2015 *Sens. and Act. B: Chem.* **207, Part B** 1053
16. Swofford H S and Carman R L 1966 *Anal. Chem.* **38** (8), 966
17. Trasatti S and Petrii O A 1992 *Jour. of Elect. Chem.* **327** (1–2), 353
18. Trucano P and Chen R 1975 *Nature* **258**, 136
19. Trijueque J et al 1999 *Electrochim. Acta.* **45,** 789